T0215827

BestMasters

Mit „BestMasters" zeichnet Springer die besten Masterarbeiten aus, die an renommierten Hochschulen in Deutschland, Österreich und der Schweiz entstanden sind. Die mit Höchstnote ausgezeichneten Arbeiten wurden durch Gutachter zur Veröffentlichung empfohlen und behandeln aktuelle Themen aus unterschiedlichen Fachgebieten der Naturwissenschaften, Psychologie, Technik und Wirtschaftswissenschaften. Die Reihe wendet sich an Praktiker und Wissenschaftler gleichermaßen und soll insbesondere auch Nachwuchswissenschaftlern Orientierung geben.

Springer awards **"BestMasters"** to the best master's theses which have been completed at renowned Universities in Germany, Austria, and Switzerland. The studies received highest marks and were recommended for publication by supervisors. They address current issues from various fields of research in natural sciences, psychology, technology, and economics. The series addresses practitioners as well as scientists and, in particular, offers guidance for early stage researchers.

More information about this series at http://www.springer.com/series/13198

Sebastian Steibl

Terrestrial Hermit Crab Populations in the Maldives

Ecology, Distribution and
Anthropogenic Impact

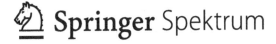

Sebastian Steibl
Animal Ecology I
University of Bayreuth
Bayreuth, Germany

ISSN 2625-3577   ISSN 2625-3615   (electronic)
BestMasters
ISBN 978-3-658-29540-0   ISBN 978-3-658-29541-7   (eBook)
https://doi.org/10.1007/978-3-658-29541-7

This Springer Spektrum imprint is published by the registered company Springer Fachmedien Wiesbaden GmbH part of Springer Nature.
The registered company address is: Abraham-Lincoln-Str. 46, 65189 Wiesbaden, Germany

Acknowledgements

Financial support for conducting the field study from Max Weber-Programm (implemented in the "Studienstiftung des deutschen Volkes" scholarship) is gratefully acknowledged. Special thanks to Naifaru Juvenile, Atoll Marine Centre and Atoll Volunteers for accommodation and support during the field study. Also, I like to thank Kuredu Resort & Spa, Mr Shahid Hussain and Komandoo Island Resort & Spa, Mr Patrick de Staercke and Hurawalhi Island Resort and Kanuhura Maldives for giving the permission to sample their islands and Mr Ray van Eeden from Prodivers Kuredu, Mr Alexis Mixe from Prodivers Komandoo, Mr Wolfgang Tippelt from Sun Dive Center Kanuhura and Ms Lisa Bauer from Marine Biology Center Hurawalhi for their contact and excellent collaboration before and during the field study. I also want to thank Korallionlab, Ms Tam Sawers and the Manta Trust charity for their advices and support during the early stages of this project. A special thanks to Mr Enrico Schwabe and the Bavarian State Collection of Zoology for helping to identify the collected gastropod species and to the whole department of Animal Ecology I, University of Bayreuth, for their scientific and personal support during this project.

Finally, I like to express my sincerest gratitude to my supervisor Prof Dr Christian Laforsch, Animal Ecology I, University of Bayreuth, for his fantastic support in planning, conducting and writing this master thesis – from the beach and from the hospital, and "from Hawaii to Maldives".

Sebastian Steibl

Table of Content

List of Figures and Tables

German Summary

Einsiedlerkrebse sind Vertreter der Ordnung Zehnfußkrebse (Decapoda). Deren auffälligstes Merkmal ist ein nicht kalzifiziertes und reduziertes Abdomen, welches durch die Behausung leerer Schneckenschalen geschützt wird. Von den über 800 Arten bewohnen nur die Vertreter der Gattung *Coenobita* den terrestrischen Lebensraum tropischer und subtropischer Küsten. Für die Küstenökosysteme haben diese Landeinsiedlerkrebse eine entscheidende Bedeutung, da sie organisches Material, wie angespülte Kadaver oder Algen, konsumieren und damit gleichzeitig ein wichtiges Bindeglied zwischen marinen und terrestrischen Nahrungsnetzen darstellen. Trotz ihrer entscheidenden Rolle für die Küstenökosysteme ist über die Biologie der Landeinsiedlerkrebse wenig bekannt. Insbesondere der Einfluss menschlicher Aktivitäten auf deren Populationen ist nahezu unerforscht. Im Rahmen der vorliegenden Masterarbeit wurde in drei Studienprojekten geklärt, (1) wie zwei Vertreter der Gattung *Coenobita* über die Ressource Schale konkurrieren, bzw. ob Konkurrenz durch Nischenbildung in Bezug auf die Gehäusewahl vermieden wird, (2) wie Landeinsiedlerkrebspopulationen auf kleinen Inselökosystemen zeitlich und räumlich verteilt sind und (3) wie verschiedene menschliche Aktivitäten einen Einfluss auf die Abundanz, Größe und Diversität der Landeinsiedlerkrebse nehmen. Die erste Studie zur Schalenpräferenz zeigte dabei, dass die beiden untersuchten Arten *Coenobita rugosus* und *C. perlatus* je eine Nische in Bezug auf Schalennutzung besetzen und somit intersezifische Konkurrenz um die limitierte Schalenressource vermeiden. Die zweite Studie zeigte, dass zeitliche und räumliche Verteilung der Landeinsiedlerkrebspopulationen äußerst variabel ist: um das Risiko des Austrocknens zu minimieren, waren die untersuchten Landeinsiedlerkrebse während der Mittagsstunden am inaktivsten, während ihre höchste Aktivität eine Stunde vor Niedrigwasser bis zum Eintreten des absoluten Niedrigwassers war. Des Weiteren bevorzugten die Landeinsiedlerkrebse Strandtypen mit einer höheren strukturellen Komplexität gegenüber reinen Sand- oder Felsstränden. Die dritte Arbeit zeigte, dass eine touristische Nutzung von Inseln negative Auswirkungen auf die Abundanz der dortigen Landeinsiedlerkrebse hat, während auf Inseln, die hauptsächlich durch Fischerei genutzt sind, kein negativer Einfluss auf die Einsiedlerkebse, dafür aber auf die Körpergröße vorhanden war. Die Ergebnisse der drei Studien liefern neue Erkenntnisse über die allgemeine Biologie der Landeinsiedlerkrebse und zeigen, wie Populationen zweier *Coenobita*-Arten durch die Interaktion mit der jeweils anderen Landeinsiedlerkrebsart, sowie durch abiotische und anthropogene Stressoren beeinflusst werden. Die Ergebnisse demonstrieren außerdem, wie variabel der Einfluss zweier unterschiedlicher Formen menschlicher Inselnutzung auf ein Inselökosystem sein kann und wie Landeinsiedlerkrebse als Werkzeug für die Evaluation eines Inselökosystems genutzt werden können.

English Summary

Hermit crabs are decapod crustaceans that are easily recognized by their uncalcified and reduced abdomen, which they protect by utilizing empty gastropod shells. Among the more than 800 hermit crab species, only a few hermit crabs, all belonging to the genus *Coenobita*, have transited into the terrestrial ecosystems and inhabit tropical and subtropical coastlines. For the coastal ecosystems, coenobitid hermit crabs play a crucial role by removing washed-up organic material and cadavers and by being the link between the marine and the terrestrial food web. Although their important role for the coastal ecosystem has been well established, little is known about their general biology. Furthermore, virtually no information is present on how different human activities are influencing hermit crab populations. Therefore, the aim of this master thesis was to investigate in three separate studies, (1) how two species of hermit crabs compete over the shell resource and if both species have established a shell preference niche, (2) how hermit crab populations are spatially and temporally distributed on small coral island ecosystems and (3) how different forms of human land use influence the overall abundance, distribution and diversity of hermit crabs. The shell preference experiments suggested that the two investigated hermit crabs, *Coenobita rugosus* and *C. perlatus*, have established their distinct niches in shell selection, thereby avoiding interspecific competition over the strictly limited shell resource. The second study revealed that the distribution of the terrestrial hermit crabs is highly variable in space and time: to minimize the risk of desiccation, hermit crabs became most inactive around midday, while being most active one hour before low tide until low tide to minimize the risk of mechanical disruption by the incoming tides. Furthermore, coenobitid hermit crabs prefer structurally more complex beach habitat types over fine sand and rock beaches. The third study showed that a touristic island use has a negative impact on the abundance of hermit crabs, while local islands, mainly used for fishery, had no impact on the abundance of hermit crabs, but on their body size. The results of the three studies offer new insights in the general biology of the coenobitid hermit crabs and show how the populations of two species are influenced by their interaction with other hermit crab species and by abiotic factors and anthropogenic stressors. The results furthermore present, how variable the impact of two different forms of human island use on an ecosystem can be and suggest that terrestrial hermit crabs, might be a useful tool in a rapid evaluation of the coastal ecosystem.

1 General Introduction

Hermit crabs (superfamily: Paguroidea) are decapod crustaceans that can be found in virtually all marine environments, from the tropics to the temperate and polar regions[1,2]. Nearly all of the more than 800 described species are characterized by an uncalcified abdominal exoskeleton[3]. Due to this unique feature, they have adapted to the utilization and occupation of empty gastropod shells (and a few other types of protective structures)[4]. By inhabiting gastropod shells, hermit crabs regain the protection, which they congenitally lack due to their uncalcified abdomen[5]. The advantageous combination of protection and mobility, which arises from the utilization of gastropod shells, is thought to be one of the key factors for the large number of hermit crabs in the marine ecosystems[4].

The strong association between hermit crabs and their utilized shells has shaped most parts of their behaviour, ecology and physiology. When the pelagic larvae metamorphoses to an adult hermit crab, they give up their symmetrical body morphology and become asymmetrical to better fit the tortuosity of the gastropod shells[5,6]. At the same time, the megalopae larvae become more benthos-associated in search for a suitable gastropod shell[7]. From the moment of their first shell utilization on, hermit crabs are in constant need for gastropod shells to grow, disperse and protect themselves and, for females, their clutch[4,8].

The selection and subsequent utilization of an empty gastropod shell is thereby not random, but an active process that selects for certain shell characteristics[9]. Hermit crabs favour gastropod shells with a sufficiently high internal volume and an adequate size, so that their body is fully protected and can be completely withdrawn into the shell[8,10]. Additionally, they select for shells with an adequate weight to optimize their locomotive activity[11,12]. In concordance, shells with a smoother surface and no long appendages are generally preferred, as spines or a rough texture might impede movement[13,14].

As empty gastropod shells with such advantageous characteristics are scarce, shell availability is thought to be always the limiting resource for hermit crab populations[15,16]. In consequence, inter- and intraspecific competition over the most suitable shell is often observed in hermit crab populations[5,17]. This forces hermit crabs to utilize gastropod shells outside their optimal shell fit range, which ultimately leads to a reduced fitness[16].

Accordingly, inhabiting a gastropod shell within the optimal shell fit range is generally associated with an increased fitness for hermit crabs: growth rate and clutch size can be maximized, while the risk of predation is minimized[8,14]. Additionally, when the utilized shell is light and well-fitting, dispersal over greater distances and foraging is optimized, as carrying a too heavy or unfitting shell

© The Editor(s) (if applicable) and The Author(s), under exclusive license to
Springer Fachmedien Wiesbaden GmbH, part of Springer Nature 2020
S. Steibl, *Terrestrial Hermit Crab Populations in the Maldives*, BestMasters,
https://doi.org/10.1007/978-3-658-29541-7_1

would waste too much energy for locomotion[18]. Utilizing gastropod shells also enabled the sea-to-land transition in the genus *Coenobita*, commonly referred to as terrestrial hermit crabs: as the physical conditions in the coastal ecosystems are harsh and bring a high risk of desiccation and disruption by wave action, the utilization of a gastropod shell minimizes evaporation and gives additional shelter from the waves, but also from terrestrial predators like sea birds[14,19]. The benefits from shell utilization has allowed the terrestrial hermit crabs to successfully establish in tropical and subtropical coastal ecosystems worldwide[20].

Due to the widespread distribution and high abundances, hermit crabs receive general awareness in the broad public, as they are well-known and easy recognizable animals on the beaches of holiday destinations throughout the world. Apart from attracting the interest of tourists while sunbathing or diving, they also play a crucial role for the environment[21]. By feeding on organic debris, terrestrial hermit crabs are important in the removal of decomposing material, like dead animal or plant material, that has been washed up on the beaches[22]. Additionally some coenobitid species are very successful in removing bird feces, as well as dead bird cadavers, from the shoreline[23]. Besides their role as beach cleaners, terrestrial hermit crabs represent an important link between marine and terrestrial food webs, as they are eaten by sea birds or bigger terrestrial crabs, thereby introducing marine biomass into the terrestrial food web[24].

However, unlike their marine counterparts, terrestrial hermit crabs have received comparably little attention in scientific research and most studies on coenobitid hermit crabs focus on the crab-shell association and their shell utilization patterns. Many other parts of the general biology of *Coenobita* species are only scarcely investigated and virtually no information can be found on how the ongoing human exploitation of the coastal ecosystems affects their populations, despite their well-established and important role for the beach ecosystem.

Therefore, the aim of this master thesis was to investigate the ecology, distribution and the anthropogenic impacts on terrestrial hermit crabs:

The first chapter aims for a better understanding of how two sympatric terrestrial hermit crab species, *Coenobita rugosus* and *C. perlatus*, coexist in their beach habitats. Combined with a field data collection, the shell preference experiments investigated, whether *C. rugosus* and *C. perlatus* compete over the shells or if both species have established their own shell niches.

In the second chapter, the temporal and spatial distribution of the terrestrial hermit crabs is described. At different day and tidal times and on the different beach habitat types, which were found on the coral islands of the Maldives, the abundances were measured.

In the third chapter, the two hermit crabs *C. rugosus* and *C. perlatus* were used as an exemplified organism to investigate, how different forms of human land use influence terrestrial hermit crabs in small coral island ecosystems. The abundance, size and diversity of the two hermit crab species and of their corresponding utilized shell was thereby compared between populations sampled on tourist islands, local islands and uninhabited islands.

In the following, the three chapters are presented as three self-contained and already peer-reviewed and published manuscripts[25–27].

2 Shell Resource Partitioning as a Mechanism of Coexistence in two Co-occurring Terrestrial Hermit Crab Species

Originally published as:
Steibl, S., & Laforsch, C. (2020) Shell resource partitioning as a mechanism of coexistence in two co-occurring terrestrial hermit crab species. *BMC Ecology, 20*(1).[25]

2.1 Abstract

Background Coexistence is enabled by ecological differentiation of the co-occurring species. One possible mechanism thereby is resource partitioning, where each species utilizes a distinct subset of the most limited resource. This resource partitioning is difficult to investigate using empirical research in nature, as only a few species are primarily limited by solely one resource, rather than a combination of multiple factors. One exception are the shell-dwelling hermit crabs, which are known to be limited under natural conditions and in suitable habitats primarily by the availability of gastropod shells. In the present study, we used two co-occurring terrestrial hermit crab species, *Coenobita rugosus* and *C. perlatus*, to investigate how resource partitioning is realized in nature and whether it could be a driver of coexistence.

Results Field sampling of eleven separated hermit crab populations showed that the two co-occurring hermit crab species inhabit the same beach habitat but utilize a distinct subset of the shell resource. Preference experiments and principal component analysis of the shell morphometric data thereby revealed that the observed utilization patterns arise out of different intrinsic preferences towards two distinct shell shapes. While *C. rugosus* displayed a preference towards a short and globose shell morphology, *C. perlatus* showed preferences towards an elongated shell morphology with narrow aperture.

Conclusion The two terrestrial hermit crab species occur in the same habitat but have evolved different preferences towards distinct subsets of the limiting shell resource. Resource partitioning might therefore be the main driver of their ecological differentiation, which ultimately allowed these co-occurring species to coexist in their environment. As the preferred shell morphology of *C. rugosus* maximizes reproductive output at the expense of protection, while the preferred shell morphology of C. *perlatus* maximizes protection against predation at the expense of reproductive output, shell resource partitioning might reflect different strategies to respond to the same set of selective pressures occurring in beach habitats. This work offers empirical support for the competitive exclusion principle-hypothesis and demonstrates that hermit crabs are an ideal model organism to investigate resource partitioning in natural populations.

2.2 Background

Throughout all ecosystems, species can be found that are closely related to each other, occupy the same trophic level within the food web and share the same habitat, thus fulfilling similar ecological roles for the ecosystem[28]. When two or more species overlap to a certain degree in their biology and share a common and essential resource that is limited in supply, these species experience competition[29,30]. This interspecific competition can occur in two forms, either via direct interference competition (i.e. fighting over resources) or via indirect exploitative competition (i.e. consumption of resources by one species makes it unavailable for second species). In ecological research, evidence for competition between two species can be provided by comparing which resources are used and which are intrinsically preferred[31].

When investigating resource utilization between co-occurring species, studies have shown that some animals that presumably compete over the same resource, actually partition the resource[32,33]. According to the competitive exclusion principle, this resource partitioning, as a form of ecological differentiation between species, can thereby be the mechanism that allows co-occurring species to coexist in the same environment[34]. This coexistence can only be realized when each species uses a discrete subset of the limiting resource, which differs qualitatively from those of the co-occurring species[35,36]. This premise for resource partitioning is described in the concept of limiting similarity, which states that there needs to be a limit to how similar two species can be to each other in order to stably coexist, rather than compete[32].

Such theoretical hypotheses are difficult to test using empirical research, as most animals in nature are not limited by only a single resource, but rather by a multitude of abiotic and biotic factors[15]. There exist, however, some co-occurring species, where enough evidence has been collected to suggest that they are indeed primarily limited by only one resource. Shell-dwelling hermit crabs are limited under natural conditions and in suitable habitats only by the availability of the shell resource, while food and habitat are not considered as a limiting factor[15,37-39]. Therefore, they appear to be suitable model organisms to investigate competition theory in empirical research.

Hermit crabs (Superfamily: Paguroidea) are characterized by an uncalcified and reduced abdomen, which they protect by utilizing mainly gastropod shells[4,40]. As a well-fitting shell optimizes growth and maximizes clutch size[8], offers protection against predators and mechanical disruption[41,42], and decreases the risk of desiccation in the intertidal and terrestrial species[13], hermit crabs are under constant pressure to find a well-fitting shell. The availability of empty and well-fitting shells thereby depends on the gastropod population and their mortality and hence is the limiting resource of hermit crab populations[4,5,15].

Co-occurring species of hermit crabs experience direct interference competition by fighting over shells in a highly ritualized behaviour and indirect exploitative competition, as the utilization of an empty shell makes it unavailable for other individuals[4,17,37,39,43,44]. This competition can force hermit crabs to utilize shells outside their optimal fit range, resulting in a reduced fitness[5,15,16]. A number of studies, however, were able to demonstrate, that, contrary to the proposed shell competition, at least some co-occurring hermit crab species partition the shell resource[15,45-47]. In these studies, the utilized gastropod shells and their morphometric parameters (e.g. size, weight) of co-occurring hermit crab species in the field were investigated and compared. It was thereby shown that co-occurring hermit crabs utilize indeed shells of different gastropod species or with different shell parameters[35,45], although other studies suggested that the observed differences in shell utilization arise not out of different preferences[37,43]. Therefore, it is discussed whether shell resource partitioning is indeed the mechanism of coexistence in co-occurring hermit crab species[15,17].

One major limitation of many research approaches that investigate shell resource partitioning in hermit crabs is that the proposed preferences are based on the species identities of the gastropod shells [e.g. 20, 26]. The utilization of different shell species depends on the gastropod communities in the particular habitat and gastropod species vary between different regions[13,16,48,49]. Proposing that co-occurring hermit crab species partition the shell resource by preferring different shell species is an uninformative and not universally applicable approach, because the available set of utilizable gastropod species varies between regions and does not reflect the actual preference of a hermit crab species, i.e. the same hermit crab species can prefer two completely different shell species in two different populations but in both cases select for the same morphological shell parameters.

A better approach is the comparison of preferences for different shell parameters. Determining the shell partitioning mechanism based on single shell parameters, however, is restricted, as the various shell variables are all highly intercorrelated, making it impossible to characterize a single parameter on which preferences could be based upon[50]. Using morphometric data, it was demonstrated that co-occurring hermit crab species have distinct preferences towards e.g. large shells or narrow apertures[45].

To deepen our understanding of resource partitioning as a possible driver of coexistence using empirical research on hermit crabs, it would be essential to incorporate (I) a large-scale sampling effort to pool data of multiple distinct hermit crab and gastropod populations, (II) a comparison between shell utilization patterns in the natural habitat and the intrinsic preferences towards distinct subsets of the resource and (III) a statistical analysis of the overall morphology of the different subsets of the resources, rather than a single parameter-approach.

The present study complies with the three abovementioned criteria by conducting an atoll-wide sampling that covered eleven distinct hermit crab and gastropod populations and by comparing the field data with laboratory shell preference experiments. A principal component analysis (PCA) of the shell morphometrics was then applied to compare the decisive criteria of the shell morphology between the co-occurring species. As research organisms to test competition theory, the only terrestrial hermit crab genus, *Coenobita*, was chosen, because it has already been established that the two co-occurring hermit crab species in the investigated system, *C. rugosus* and *C. perlatus*, are both primarily beach associated and unspecialized detritus feeders with no clear food preferences[21,51,52]. They are therefore an ideal system to test for the effect of the shell resource on coexistence, because other potentially limiting factors can be excluded upfront. The overall shell utilization in land hermit crabs has received only limited research focus in comparison to their well-studied marine counterparts[53,54]. As terrestrial hermit crabs are restricted to the island, they inhabit and obtain the shell resource only from the surrounding coastal water[13]. Therefore, sampling multiple islands covers distinct hermit crab and gastropod populations and decreases the effect of predominant species in one island ecosystem.

2.3   Results

2.3.1   Field Data

Of the 876 collected hermit crabs, 700 were identified as *C. rugosus* and 176 as *C. perlatus*. The proportion of *C. rugosus* and *C. perlatus* varied significantly between the eleven investigated islands ($F = 6.2536$, $df = 10$, $P < 0.001$). On nine out of the eleven investigated islands within the Atoll, the mean proportion of *C. rugosus* was $86.47 \pm 11.64\%$. On one island however, only 37.05% of the collected crabs were identified as *C. rugosus*, while 62.95% were *C. perlatus*. On another island, *C. perlatus* was completely absent from the investigated plots.

The proportion of *C. rugosus* ($80.28 \pm 7.10\%$) and *C. perlatus* ($19.72 \pm 7.10\%$) was not significantly different between the four investigated beach habitat types ($F = 1.9196$, $df = 3$, $P = 0.147$). The collected *C. rugosus* and *C. perlatus* had a carapace length of 6.50 ± 2.23 mm and 6.46 ± 2.71 mm, respectively. The mean carapace length of the two species did not differ statistically (Wilcoxon W = 56344, $P = 0.291$). The collected *C. rugosus* inhabited gastropod shells of 90 different species (in 21 different families), while the collected *C. perlatus* inhabited gastropod shells of 41 different species (in 14 different families). The shell species diversity index, i.e. the diversity of shell species inhabited by the two investigated hermit crab species, of *C. rugosus* was H = 3.644 and of *C. perlatus* H = 3.039. The niche width in respect to utilizable shell species was therefore B = 23.870 for *C. rugosus* and B = 12.869 for *C. perlatus* (Tab. 1).

The proportional utilization of the investigated shell types differed significantly between *C. rugosus* and *C. perlatus* (Tab. 1). Proportionally more *C. rugosus* inhabited naticid shells than *C. perlatus* ($P = 0.003$), while proportionally more *C. perlatus* inhabited cerithiid ($P < 0.001$) and strombid shells ($P < 0.001$). No differences were found in the number of inhabited nassariid shells between *C. rugosus* and *C. perlatus* ($P = 0.237$; Tab. 1).

Table 1: Comparison of the shell utilization and preferences of the two co-occurring hermit crab species. Asterisks (*** $P < 0.001$) indicate significant differences in the proportional utilization or selection of the respective shell type between the two hermit crab species, C. rugosus and C. perlatus.

	Coenobita rugosus	*Coenobita perlatus*
Utilized gastropod shells	90 species (21 families)	41 species (14 families)
cerithiid shells utilized	13.90%	32.06% (***)
cerithiid shells selected	54.67%	56.00%
nassariid shells utilized	28.78%	18.49%
nassariid shells selected	64.00%	65.33%
naticid shells utilized	14.09%	4.22% (***)
naticid shells selected	56.00%	20.00% (***)
strombid shells utilized	12.77%	39.52% (***)
strombid shells selected	25.33%	58.67% (***)
Shell diversity Shannon H	3.644	3.039
Niche width B of shell species	23.870	12.869

2.3.2   Shell Preference Experiments

The mean carapace length of the 150 tested *C. rugosus* was 6.25 ± 1.43 mm and of the 150 tested *C. perlatus* 6.42 ± 1.42 mm (mean ± standard deviation). The size of the tested hermit crab in the laboratory experiment did not differ statistically between the two species (Wilcoxon W = 12207, P = 0.199).

The two terrestrial hermit crabs *C. rugosus* and *C. perlatus* had significantly different shell preferences for the tested gastropod shells (Tab. 1). *C. perlatus* selected strombid shells significantly more often than *C. rugosus* (P < 0.001) and *C. rugosus* selected naticid significantly more often than *C. perlatus* (P < 0.001). No differences existed for the number of selected cerithiid (P = 1.000) and nassariid shells (P = 1.000) between the two hermit crab species.

2.3.3   Morphometric Analysis of Gastropod Shells

The five investigated morphometric parameters (shell length, shell width, aperture length, aperture width, shell weight) of the utilized gastropod shells differed significantly between the four investigated gastropod shell types (F = 71.505, df = 3, P < 0.001) and between the two hermit crab species (F = 16.080, df = 1, P < 0.001).

The first three principal components of the PCA, comparing the morphometric parameters, explained 96.47% of the total variance and were therefore used for further analysis (Fig. 1). Principal component 1 (PC1) correlate with all five morphometric parameters, suggesting that all five parameters vary together. PC2 is primarily a measure for shell length (correlation 0.784) and aperture width (correlation -0.526) and can be viewed as an overall descriptor of the shell shape with high values of PC1 indicating an elongated and narrow shell shape, while low values of PC2 indicate a short and bulbous shell shape. PC3 negatively correlates with aperture length (correlation -0.851) and can be viewed as a measure of how elongated the shell aperture is (Tab. 2).

The four gastropod shell types differed significantly in PC1 (F = 60.96, df = 3, P < 0.001), PC2 (F = 548.1, df = 3, P < 0.001) and PC3 (F = 307.8, df = 3, P < 0.001). Tukey HSD post-hoc test indicated significant differences in PC1 between all pairwise comparisons (P < 0.001), apart from nassariid-cerithiid (P = 0.997) and strombid-naticid shells (P = 0.999). PC2 was significantly different in all pairwise comparisons (P < 0.001 in all comparisons). PC3 was significantly different in all comparisons (P < 0.001), apart from one non-significant difference in the pairwise comparison of nassariid and cerithiid shells (P = 0.051; Tab. 2).

All three principal components of the shell parameters differed significantly between the two hermit crab species (PC1: F = 9.819.3, df = 1, P = 0.001; PC2: F = 57.01, df = 1, P < 0.001; PC3: F = 92.14 df = 1, P < 0.001).

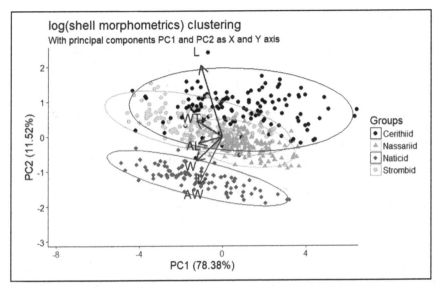

Figure 1: The shell morphology of the four most utilized gastropod shell types. The principal component analysis is based on the five log-transformed morphometric parameters (*AL* aperture length, *AW* aperture width, *L* length, *W* width, *WT* weight). Each data point represents a single shell, colours resemble the different shell types.

Table 2: Comparison of the shell morphology of the four most utilized gastropod shell types and the two hermit crab species. Principal components (PC) of the PCA are based on five morphometric parameters of the four utilized gastropod shell types. Significant differences between the mean PC values for each shell type are indicated by different letters behind the PC value, same letters indicate no statistical difference between the PC values of the respective shell types.

	PC 1	PC 2	PC 3
shell length	-0.396	0.784	0.080
shell width	-0.485	-0.265	0.016
aperture length	-0.438	-0.078	-0.851
aperture width	-0.437	-0.526	0.362
shell weight	-0.472	0.174	0.370
Shell thickness	-0.329	-0.804	0.046
cerithiid shells	0.874 (A)	0.765 (A)	0.372 (A)
nassariid shells	0.839 (A)	-0.200 (B)	0.268 (A)
naticid shells	-1.198 (B)	-1.189 (C)	0.056 (B)
strombid shells	-1.195 (B)	0.384 (D)	-0.791 (C)
Coenobita rugosus	0.151 (A)	-0.134 (A)	0.046 (A)
Coenobita perlatus	-0.479 (B)	0.424 (B)	-0.146 (B)

2.4 Discussion

According to the competitive exclusion principle, ecological differentiation is the premise for coexistence in co-occurring species[34]. This ecological differentiation can be realized by partitioning the limiting resource between two species[36]. In the present study, the utilization of the limiting resource of two co-occurring hermit crab species was investigated to study the relevance of resource partitioning as a driver of coexistence. In natural populations, the two co-occurring hermit crabs *C. rugosus* and *C. perlatus* utilized different gastropod shell species. These differences in the shell utilization of the two hermit crab species arise out of different preferences towards different shell types. Together with the morphometric analysis, the presented data suggest that the two hermit crab species are not in competition over the limited shell resource but have evolved different preferences towards distinct subsets of the shell resource, which ultimately could enable both species to coexist in their habitat.

Coexistence of co-occurring marine hermit crabs has been suggested to arise out of a combination of resource and habitat partitioning[4,15]. Terrestrial hermit crabs are more restricted in their habitat choice, as especially small islands offer only little heterogeneity in the beach environment[19,20,55,56]. Although *C. perlatus* was overall less abundant than *C. rugosus*, there relative proportions did not differ between the four present beach habitat types. As both species are known to be primarily beach-associated and not occurring in the densely vegetated inland[24,57–60], the high overlap of both species in the beach habitats suggests that habitat partitioning is not a driver of coexistence in these two species.

Partitioning of or competition over the food resource can also be excluded as a driver for coexistence, as previous studies demonstrated that *C. rugosus* and *C. perlatus* are both unspecific detritus feeders with no clear food preference[52,59] and not limited by food availability[4,15,44].

As habitat and food resource partitioning appears to play a minor role for *C. rugosus* and *C. perlatus*, the possible mechanism for coexistence might arise out of shell resource partitioning. The morphometric analysis of the utilized shells in the field suggest that *C. rugosus* utilizes shells with a small and globose morphology, while *C. perlatus* utilizes shells with a large, elongated and narrow morphology. These utilization patterns arise indeed out of different intrinsic preferences towards the respective shell morphology, as *C. rugosus* selected for the short and globose naticid shells, while *C. perlatus* selected for the large and elongated strombid shells in the laboratory experiments. The determined preferences towards a certain shell morphology lay in concordance with previous studies, which reported *C. rugosus* to utilize mainly Muricidae, Neritidae or Turbinidae shells, which also have a globose morphology, and *C. perlatus* to utilize mainly the elongated cerithiid shells[54,57,59–61]. This overall similarity further underlines that not the shell species itself is the decisive criteria in the shell selection process, but rather the overall morphology of the present shell, described by the principal components of the morphometric data. The utilized shells found in the natural populations were overall fairly eroded and showed no striking variations in colour or ornamentation but appeared rather uniform pale and smooth, independent of the gastropod species. Therefore, preferences towards certain shell colours or ornamental features like spines can be excluded as further decisive factors in shell selection of the investigated hermit crab species. As gastropod communities vary between different regions, the adaptive mechanism in shell selection behaviour is therefore not the evolution of preferences towards species (although at least one hermit crab species is known utilizing only one shell species, *Calcinus seurati*[4,5]), but rather of preferences towards certain shell morphologies[62].

The two investigated hermit crab species apparently have evolved different shell preferences towards distinct subsets of the shell resource. These intrinsic preferences could hint towards differing strategies of the two hermit crab species to respond to the same overall selective pressures[9,63]. Heavy and elongated shells with a narrow aperture, like the strombid shells, offer optimal protection against desiccation and predation, but limit clutch size and increase energy expenditure during locomotion due to a reduced internal volume and increased weight[5,8,35,45]. Light-weight and voluminous shells, like the naticid shells, allow a greater dispersal and are advantageous for burrowing, but cannot retain water efficiently and offer less protection against predation[47,57,64]. As different shell preferences might represent different strategies to respond to selective pressures from the same environment, *C. perlatus* might has evolved a strategy to reduce desiccation- and predation-related mortality at the expense of an increased energy expenditure and limited clutch size[9]. *C. rugosus* has evolved a strategy to maximize reproductive output at the expense of an increased susceptibility for desiccation and predation.

Further research is needed to test, whether the observed shell resource partitioning in the two co-occurring hermit crab species is the cause or the effect of the proposed ecological differentiation in respect to their life-history strategy and if the utilization of different subsets of the shell resource can even be a driver of speciation in hermit crabs.

In either way, it is shown that the utilization of distinct subsets of the limiting resource can drive ecological differentiation, which then ultimately enables two species to coexist[34,36]. It is thereby demonstrated that co-occurring hermit crabs are indeed suitable model organisms to empirically investigate competition and coexistence theory, as their limitation by primarily one resource offers controllable and empirically testable conditions for investigating natural and intrinsic behaviour of resource partitioning.

2.5   Conclusion

Overall, our research investigated the mechanism of resource partitioning as a driver of coexistence and demonstrated that two co-occurring species of terrestrial hermit crabs have evolved intrinsic preferences towards distinct subsets of the shell resource, which attenuates interspecific competition over the limiting resource in natural populations. As the preferred shell morphologies of the two hermit crab species either maximize reproductive output or minimize predation risk, the two hermit crab species might have evolved different strategies to respond to the overall selective pressures in their natural habitat.

These findings offer empirical support for theoretical hypotheses on competition theory and mechanisms of coexistence in ecology. By discussing different life-history strategies, associated with the observed resource partitioning, the presented model system using hermit crabs can form the basis for future research on mechanisms of coexistence and speciation.

2.6   Methods

2.6.1   Field Data

Hermit crabs were collected on the beaches of eleven coral islands, distributed over the Lhaviyani (Faadhippolhu) Atoll, Republic of Maldives. Sampling was carried out between 03/02/2017 and 10/03/2017, always in the time from two hours before low tide until absolute low tide. On each island, hermit crabs were collected in six plots with 10 m length (measured along the current drift line) and 2 m width (measured perpendicular to the current drift line). The habitat structure of each plot was assigned in four different beach habitat types: (1) fine sand beach, (2) fine sand beach interspersed with small coral and rock fragments, (3) fine sand beach interspersed with larger boulders and (4) predominantly rock-covered beach. The collected hermit crabs were transferred to the laboratory and removed from their shell by carefully heating the apex of the shell above an open flame. This is a standard procedure when investigating hermit crabs and leaves the animal without injuries[47,64]. Afterwards, the hermit crab and their corresponding shell were photographed on millimetre paper (Nikon D5000 mounted with Nikon AF-S Nikkor 18-105 mm, 1:3.5-5.6, Nikon Corp., Tokyo, Japan.) and identified using identification keys[65–69]. The weight of the shell was measured using a fine scale (TS-300 300g x 0.01g, G&G GmbH, Neuss, Germany).

The carapace length of the hermit crabs and the morphometric parameters of their corresponding shell were determined using ImageJ 1.49b (Rasband, W.S., ImageJ, U. S. National Institutes of Health, Bethesda, Maryland, USA, http://imagej.nih.gov/ij/, 1997-2015). Shell length was measured from the shell's apex to the siphonal notch - if present - or otherwise to the lower end of the aperture. Shell width was measured perpendicular to the longitudinal axis of the shell at the broadest section. Shell aperture length was measured from the anterior to the posterior canal of the aperture and aperture width was measured perpendicular to the aperture length between the outer lip and the columellar fold at the broadest section.

Statistical analysis was performed using R 3.5.1.[70] Differences in the number of shells utilized for a given shell species between *C. rugosus* and *C. perlatus* were tested for the four most abundant gastropod families in the plots, i.e. strombid shells (246 specimen), nassariid shells (196 specimen), cerithiid shells (166 specimen) and naticid shells (141 specimen; Fig. 2). Statistical comparison in the number of utilized shells of each of the four shell types between the two collected hermit crab species were analysed using Fisher's exact test[71]. Levels of significance were adjusted using Bonferroni-Holm-correction. The relative abundance of the two hermit crab species was calculated and statistically compared between the four investigated beach habitat type and between the eleven investigated coral islands using non-parametric multivariate analysis (PERMANOVA) with 999 permutations, implemented in the vegan package of R[72]. The diversity of shell species occupied by the two hermit crab species was calculated using the Shannon-Index H. Based on the number of inhabited shells from the two hermit crab species, the niche breadth (B) with respect to shell species inhabited was calculated using

$$B = \frac{1}{\Sigma(p_i{}^2)}$$

where p_i is the proportion of crabs (*C. rugosus* or *C. perlatus*) found in shells of the gastropod species I[39]. The sizes of the two sampled hermit crab species were statistically compared using Wilcoxon test.

2.6.2   Shell Preference Experiments

150 hermit crabs of each of the two species *C. rugosus* and *C. perlatus* and 150 cerithiid, nassariid, naticid and strombid shells were collected on the beaches of Naifaru, Lhaviyani (Faadhippolhu) Atoll, Republic of Maldives from 16/03 to 20/03/2017. The collected hermit crabs were transferred into the laboratory and removed from their shell. After removing the crab out of its shell, the carapace length was measured using a ruler and the size of the crab with its corresponding shell was noted.

One hermit crab (without its shell) of a given size was then transferred into a 45-cm diameter test arena, filled 2 cm with sand from the adjacent beaches, and left to acclimatize for 5 min. After acclimatisation, two of the four tested shell types, were placed next to each other on a random place inside the test arena with the aperture facing upwards. For each tested hermit crab of a given size, two empty gastropod shells were presented that were formerly inhabited by a hermit crab with the same size of the one tested in the arena (e.g. a 1 cm-sized hermit crab was offered two shells that were formerly inhabited by 1 cm-sized crabs). This procedure was conducted to ensure that both presented shells were principally utilizable for the tested hermit crab of a given size. For *C. rugosus* and *C. perlatus* each combination of two shell species (strombid vs. naticid., strombid. vs. nassariid., strombid. vs. cerithiid., naticid vs. nassariid, naticid vs. cerithiid, nassariid vs. cerithiid) was tested 25 times ($N = 25$). One hour after presenting the two empty gastropod shells, the utilized shell type was noted and the hermit crab together with both shells transferred back to its original habitat. If no shell had been utilized by the tested hermit crab after one hour, the experiment was terminated and the crab, as well as both shells, excluded from the experiment and transferred back to the original habitat.

The carapace lengths between the two tested hermit crab species was statistically compared using the Wilcoxon test. Preferences for the investigated shell species, between the two hermit crab species were analysed using Fisher's exact test. Levels of significance were adjusted using Bonferroni-Holm-correction.

2.6.3   Morphometric Analysis of Gastropod Shells

Differences in the five morphometric parameters between the four different gastropod types and the two hermit crab species were compared using non-parametric multivariate analysis (PERMANOVA) with 999 permutations. One principal component analysis (PCA) was performed with log-transformed values of the five morphometric parameters. Statistical differences between the principal components of the four shell types and the two hermit crab species were analysed using ANOVA and Tukey HSD post-hoc tests.

Figure 2: The two co-occurring hermit crab species and the four most commonly utilized gastropod shell types. On the top, the two tested hermit crab species, *Coenobita rugosus* (**a**) and *C. perlatus* (**b**) and below the four different shell types utilized, i.e. nassariid (**c**; here depicted: *Nassarius variciferus*), naticid (**d**; here depicted *Polinices mammilla*), cerithiid (**e**; here depicted *Rhinoclavis aspera*) and strombid shells (**f**; here depicted *Gibberulus gibberulus*)

3 Daytime Activity and Habitat Preferences of two Sympatric Hermit Crab Species (Decapoda: Anomura: *Coenobita*)

Originally published as:
Steibl, S., & Laforsch, C. (2019) Daytime activity and habitat preferences of two sympatric hermit crab species (Decapoda: Anomura: *Coenobita*). *Estuarine, Coastal and Shelf Science, 231*(106482).[26]

3.1 Abstract

The beach environment is extremely dynamic in space and time. Abiotic factors like tides, sun exposure or sediment structure are defining the ecology of the beach-associated fauna. Among the most common beach-dwelling organisms of tropical and subtropical shores are the hermit crabs of the genus *Coenobita*[73] (Decapoda: Anomura). They utilize gastropod shells to protect against predators, to avoid desiccation and disruption by wave action and further show behavioural adaptations, like burrowing in the substratum to withstand the abiotic stressors of coasts. Little is known, however, if the abiotic factors of the beach habitat influence the daytime activity and habitat preferences. We therefore analysed the changes in abundance during daytime, at different tidal times and in different coastal habitats in a community of two sympatric *Coenobita* species, *C. rugosus* and *C. perlatus*. We hypothesized that habitat, daytime and tidal time influenced the overall abundance. Here, we showed that hermit crabs became largely absent during midday, while their highest diurnal activity laid in the two hours before low tide until absolute low tide. Structurally more complex beach types were preferred over pure fine sand or rock beaches. These behaviours and preferences of the investigated hermit crabs are adaptive as they aid in avoiding desiccation, while becoming most active when food availability is highest during low tide. Heterogenous beach habitats are probably favoured over homogenous sandy beaches, because accumulation of marine debris, a major food source, is increased. This emphasizes, how physically controlled the distribution of beach-dwelling organisms is and demonstrates how abiotic stressors can become major drivers for behavioural adaptations in beach crustaceans.

3.2   Introduction

Beaches are among the most dynamic terrestrial habitats[74]. They are shaped by the temporal and spatial variations in tidal regime, wave climate, sun exposure, heat and different sediment types[75]. The combination of these parameters causes a wide range of morphodynamic beach types and physically structures the beach macrofauna[76,77]. Fine sand beaches have the overall lowest species diversity, while a higher substrate complexity is linked to a greater species diversity and also to higher abundance of organisms[78,79]. These variations in the overall species richness, density and abundance between different beach communities are thereby caused by the individual responses of the beach-dwelling taxa to the physical parameters[80].

To adapt to the variable beach environment, beach-dwelling organisms have evolved various behavioural and physiological mechanisms to withstand the selective pressure of the physical factors they experience on the beach habitat[81]. The orientation and navigation along physical gradients towards beneficial or away from disadvantageous conditions in the beach environment is widespread in many beach-associated organisms[82]. A circadian rhythm with either diurnal, nocturnal or crepuscular activity peaks aids in avoiding unfavourable conditions, e.g. heat during the day or predation pressure during night[83,84]. The synchronisation of activity patterns to the tidal regime always keeps the animals in an area with optimal feeding conditions and decreases the risk of displacement by heavy wave action[85,86]. Similarly, many beach-associated crustaceans synchronise their movement to the intensity of wave action to stay in the zone of greatest water movement, where food availability is maximised[83].

A common and ubiquitous organism in the tropic and sub-tropic coastal habitats are the terrestrial hermit crabs (Decapoda: Anomura: Coenobitidae). They belong to the genus Coenobita[73] and comprise 25 species[87]. Besides predation, coenobitid hermit crabs utilize gastropod shells to minimize evaporation and the risk of desiccation in order to persist in the beach environment[58]. They furthermore display behavioural adaptations to withstand the unfavourable and harsh physical conditions of the beach environment[20]. During the day, they burrow themselves to avoid direct sun, dehydration and increase in body temperature, making them mainly active at night and during dawn[57,88]. Vertical movement with tidal times on the beach habitat is displayed to avoid displacement during feeding on washed-up material at the drift line[88]. Because organic debris, like seagrasses or algae, is the main food source for coenobitid crabs, they accumulate and cluster in areas of the beach environment, where the amount of detritus is high[21,89].

Although previous studies report clustering behaviour in coenobitid hermit crabs and suggest an overall night-active life habit that follows the tidal regime, many of the physical parameters shaping the temporal and spatial distribution of coenobitid hermit crabs in the beach environment are not thoroughly understood. Page & Willason (1982) reported for example that the coenobitid hermit crab *C. rugosus*[90] are commonly found during day and night, while other studies attribute a strict nocturnal activity to *C. rugosus*[58,88,91]. Barnes (1997) observed no dependence of the activity of *C. rugosus* on tidal activity, while other studies suggest a strong influence of the tidal regime on coenobitid abundances[4]. Additionally, little is known if and how the beach habitat type itself shapes the spatial distribution of coenobitid hermit crabs.

The aim of this study was therefore, to investigate how habitat characteristics together with daytime and tidal time influence distribution of coenobitid hermit crabs in space and time. In the first experiment, the spatial distribution of hermit crab abundance on different types of beach habitats, ranging from a fine sand beach to a predominantly rock-covered beach, was measured. In a second experiment, samplings from sunrise to sunset with changing tidal times were carried out in the same plot to identify when the highest and lowest abundances of hermit crabs occur throughout a day.

3.3   Material and Methods

The investigated hermit crab community in the studied system comprised two species, *Coenobita rugosus* and *C. perlatus* [92]. The two species occurred in a fairly constant ratio of 8:1 (*C. rugosus* - *C. perlatus*) throughout the whole investigated atoll[25]. The mean body size (measured as shield length) of the hermit crabs in the investigated system was 0.62 ± 0.19 cm (mean $\pm$ SD) for *C. rugosus* and 0.61 ± 0.21 cm for *C. perlatus*[25].

3.3.1   Habitat Preferences of a Coenobitid Hermit Crab Community

Sampling was carried out on 07/02/17 and 08/02/17 during falling tides between 7:17 am and 12:42 pm on Naifaru Island, Lhaviyani Atoll, Republic of Maldives. The spatial variability of hermit crab distribution was recorded by assigning the beach into four categories: (1) fine sand (FS), (2) fine sand with small fragments (SF), (3) fine sand with larger rocks (LR) and (4) predominantly rock-covered beach (RC). Each of the four beach types was replicated six times by sampling different locations along the whole shoreline of the investigated island ($N = 6$). The investigated plots measured 10 m along the current drift line and 2 m landwards, measured from the current drift. Each plot was measured once and the total number of hermit crabs within each plot was recorded. For statistical analysis, hermit crab density of the hermit crab community per plot, i.e. 20 m², was calculated. Differences in the hermit crab density between the four beach habitat types were statistically compared with Kruskal-Wallis test and Dunn-Bonferroni post-hoc tests and Bonferroni corrections using R 3.3.0.

3.3.2   Daytime Activity of a Coenobitid Hermit Crab Community

Sampling was carried out on nine days in the time between 12/01/17 and 05/02/17 on Naifaru Island, Lhaviyani Atoll, Republic of Maldives (Tab. 3). The temporal variability of hermit crab distribution was investigated in a single plot that comprised fine sand and fine sand with small fragments habitats. The plot was 25 m long (measured along the drift line) and 2 m wide (measured perpendicular to the drift line). The vertical position of the plot was adjusted hourly to assure that the plot always covered the first 2 m landwards from the current drift line. The exact start and end position of the plot was marked using GPS (eTrex Vista® Cx, Garmin Ltd., Schaffhausen, Switzerland). All hermit crabs in the plot were counted every hour from sunrise to sunset with the exact hourly sampling times adjusted for every sampling day individually, based on the time of absolute low tide event for that day. For example, when absolute low tide occurred at 8:36 (e.g. sampling day six), sampling was conducted at 6:36, 7:36, 8:36, etc. (Tab. 3). To analyse the temporal variation in abundance, the mean abundance for each sampling during falling tides (i.e. the hourly measured abundance from two hours before low tide until absolute low tide) and during rising tides (i.e. the hourly measured abundance in the first three hours after absolute low tide) was calculated and statistically compared with a non-parametric Kruskal-Wallis test.

Table 3:   Tidal events, sunrise and sunset for the nine sampling days on Naifaru, Republic of Maldives. Low tide events in bold mark the time to which each sampling day was adjusted

date	sunrise	sunset	high tide	low tide
12/01/17	6:20	18:09	0:27, 11:26	**6:34**, 17:58
23/01/17	6:22	18:14	23:17	**15:04**
24/01/17	6:22	18:14	8:51	5:54, **16:05**
25/01/17	6:22	18:15	10:08	6:09, **16:50**
26/01/17	6:22	18:15	0:15, 11:01	**6:29**, 17:31
31/01/17	6:23	18:16	2:19, 14:15	**8:36**, 20:14
01/02/17	6:23	18:17	2:42, 14:58	**9:05**, 20:47
02/02/17	6:23	18:17	3:07, 15:45	**9:40**, 21:20
05/02/17	6:23	18:17	4:28, 19:59	**12:05**

3.4 Results

3.4.1 Habitat Preferences of a Coenobitid Hermit Crab Community

Hermit crab density differed significantly between the four beach habitats (Kruskal-Wallis: $df = 3$, $\chi^2 = 17.739$, $P < 0.001$; Fig. 3). Density was significantly higher in the 'fine sand with small fragments' beach habitat than in the 'fine sand' habitat (Kruskal-Dunn post-hoc: $P = 0.006$) and in the 'predominantly rock-covered' habitat (Kruskal-Dunn post-hoc: $P = 0.007$). Density was significantly lower in the 'fine sand' habitat than in the 'fine sand with larger rocks' habitat (Kruskal-Dunn post-hoc: $P = 0.050$). Hermit crab density did not differ between the 'fine sand' habitat and the 'predominantly rock-covered' habitat (Kruskal-Dunn post-hoc: $P = 1.000$) and the 'fine sand with larger rocks' habitat (Kruskal-Dunn post-hoc: $P = 0.057$). The density did not differ between the 'fine sand with larger rocks' habitat and the 'fine sand with small fragments' habitat (Kruskal-Dunn post-hoc: $P = 1.000$).

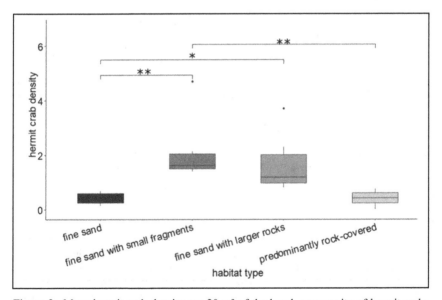

Figure 3: Mean hermit crab density per 20 m² of the beach community of hermit crabs *Coenobita rugosus* and *C. perlatus* in the four different beach habitat types. *$P < 0.05$; **$P < 0.01$ (Kruskal-Wallis: $\chi^2 = 17.739$).

3.4.2   Daytime Activity of a Coenobitid Hermit Crab Community

The abundance of *C. rugosus* and *C. perlatus* varied throughout the day (Fig. 4). Independently of the low tide event of each sampling day, the first abundance minimum always occurred around midday. A second abundance minimum appeared in the late afternoon, shortly before sunset (between 18:09 and 18:17). The abundance maximum of each sampling day occurred either in the morning or in the afternoon and laid in the two hours before low tide until absolute low tide. Only on one sampling day, where low tide occurred directly at midday (12:05), the abundance maximum occurred four hours earlier (08:05). A diurnal abundance maximum was never observed after absolute low tide during rising tides. The abundance of hermit crabs was significantly higher during falling tides than during rising tides (Kruskal-Wallis: $\chi^2 = 19.000$, $P < 0.001$; Fig. 5).

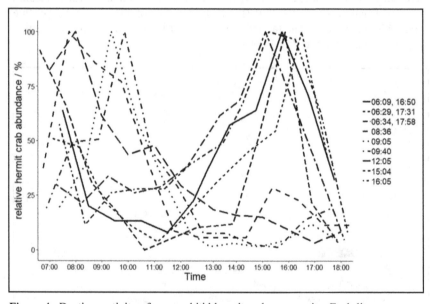

Figure 4: Daytime activity of a coenobitid hermit crab community. Each line represents an individual sampling day ($N = 9$), measured hourly from sunrise to sunset. For each sampling day, the abundance was calculated relative to the daily maximum. The low tide event for each sampling day is given on the legend to the right (see also tab. 3 for low and high tide events).

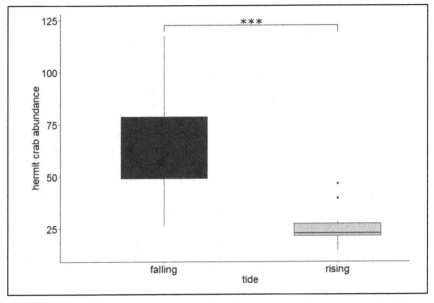

Figure 5: Mean abundance of the beach community of hermit crabs *Coenobita rugosus* and *C. perlatus* during falling and rising tides ($N = 10$). ***$P < 0.001$ (Kruskal- Wallis, χ^2 = 19.000).

3.5 Discussion

The results of the present study showed that coenobitid hermit crabs are strongly influenced in their diurnal activity by the structural habitat characteristics, day-time and tidal time. On the same island, significant variations in the abundance between different beach types demonstrated that coenobitid hermit crabs are more abundant in a more structured and complex habitat. At the same time, the abundance is highest before absolute low tide, while activity is, independently of tidal time, always decreased around midday.

Diurnal activity of coenobitid hermit crabs is influenced by variations in humidity[14]. As coenobitid crabs normally burrow themselves during the day to avoid desiccation and to gain water from the sand, the relative high humidity in the tropical Maldives of about 80% in average[93] could explain the increased activity during day in the present study, as the necessity to avoid the daily heat is reduced[94]. In concordance with Barnes (1997), an overall abundance low was observed around midday, independently of the tidal regime, probably to avoid desiccation at the hottest time of the day.

A dependence on the tidal regime of coenobitid hermit crabs was nevertheless shown by Vannini (1976), Grubb (1971) and Hazlett (1981), although only observed during night. In contrast, Barnes (1997) observed no dependence on tidal times in *C. rugosus*. Unlike the aforementioned studies, the present study showed that there also exists a dependence on tides during daytime activity. This suggests that spatial and temporal variations in coenobitid hermit crab activity arise out of a combination of an endogenous circatidal and circadian rhythm component[95].

Synchronising activity to the tides is adaptive for coenobitid hermit crabs, as the risk of being displaced by wave action is diminished during falling tides and around absolute low tide[85]. Coenobitid hermit crabs are feeding on washed-up organic material at the drift line, thereby exposing themselves to the risk of being caught by waves[23,96]. During rising tides and at high tide, this risk of displacement is strongly increased[82], which explains the measured abundance decline after low tide events in the present study. The overall input of accumulated washed-up organic material on the beaches is also influenced by the tidal regime [97]. An activity pattern, synchronised to falling tides, is also shown in many other beach-dwelling organisms, which wait until conditions become favourable to emerge for feeding[98].

The overall high abundance of coenobitid crabs during day (in average 27 individuals with peaks of 119 coenobitid crabs in the plot, i.e. per 50 m²) are contradictory to most studies suggesting a nocturnal activity, but an inactivity during day[21,58,88,91]. The reasons for the observed diurnal activity pattern in the present study might arise out of a sampling procedure with an overall higher temporal resolution, as the hermit crabs were counted every hour, while e.g. Page & Willason (1982) only sampled every four hours, thereby reducing the possibility to capture diurnal variations in the abundances[21]. Vannini (1976) and Grubb (1971) observed abundance peaks of *C. rugosus* between one to two hours before sunset, while in present study the abundance was high in the late afternoon (around 16:00), but then halved every hour until it reached a minimum around sunset[60,91]. *C. rugosus,* however, was observed to be nocturnal in some studies[4], while others have observed similar activities at night and day[21]. This could indicate differences in the circadian activity patterns, depending on the geographical location or physiological condition of the animals[21,99,100].

Besides the changing physical conditions over the course of a day, the habitat itself also had a great influence on the activity of beach-dwelling coenobitid crabs. The significantly higher abundance in the 'fine sand with small fragments' and 'fine sand with larger rock' habitats compared to the less heterogenous 'fine sand' and 'rock-covered' habitats might result from a greater food availability in these structurally more complex and heterogenous beach habitat types[59,101–103]. An artificial increase in the amount of rocks, logs, coconut husks and other detritus on the same beach location was shown to double the hermit crab abundance within one day[21]. The coral fragments and rocks in the investigated beach habitats in the present study may facilitate that a higher amount of organic material is detained from being rapidly flushed away by the waves[101]. Hence, these structurally more complex habitats are likely favoured by the coenobitid hermit crabs as overall food availability is increased[89,103,104]. Besides the increased food availability, it is also likely that more empty gastropod shells, which get washed ashore, accumulate in these more heterogeneous beach habitat types or get translocated into this habitat by foraging hermit crabs[105]. As gastropod shells are the limiting resource for hermit crab populations, areas, where shell accumulation is increased, could show increased hermit crab abundance[15].

The results of the present study demonstrate, how the physical parameters of the beach environment shape the diurnal and spatial distribution of coenobitid hermit crabs. This shows that besides their shell utilization behaviour, hermit crabs display further behavioural responses to withstand the environmental conditions of beaches. This study thereby confirms that the harsh abiotic factors of beaches are major drivers for behavioural adaptations in beach-dwelling organisms[82].

4 Disentangling the Environmental Impact of Different Human Disturbances: a Case Study on Islands

Originally published as:
Steibl, S., & Laforsch, C. (2019) Disentangling the environmental impact of different human disturbances: a case study on islands. *Scientific Reports, 9*(13712).[27]

4.1 Abstract

Coastal ecosystems suffer substantially from the worldwide population growth and its increasing land demands. A common approach to investigate anthropogenic disturbance in coastal ecosystems is to compare urbanized areas with unaffected control sites. However, the question remains whether different types of anthropogenic disturbance that are elements of an urbanized area have the same impact on beach ecosystems. By investigating small islands that are utilized for tourism, inhabited by the local population, or remained completely uninhabited, we disentangled different anthropogenic disturbances and analysed their impacts on hermit crabs as indicator species. We observed a negative impact on abundance on tourist islands and a negative impact on body size on local islands. In comparison to the uninhabited reference, both disturbances had an overall negative impact. As both forms of disturbance also impacted the underlying food resource and habitat availability differently, we propose that the findings from our study approach are valid for most obligate beach species in the same system. This demonstrates that in urbanized areas, the coastal ecosystem is not always impacted identically, which emphasizes the importance of considering the particular type of anthropogenic disturbance when planning conservation action in urbanized areas.

4.2 Introduction

Our planet faces an ever increasing number of environmental problems caused by the growth of the human population and its land demands[106]. One ecosystem that suffers substantially from population growth are coasts. Between 50% and

75% of the world's population live close to coasts[107], thereby intensifying the anthropogenic impacts on this fragile environment. Globally, sand-dominated beaches comprise 75% of the ice-free coastline[108] – and in addition to their inherent ecological value, they form a crucial component of the travel and tourism industries worldwide[109].

Many ecological studies try to identify factors that impact sandy beach ecosystems for the development of conservation measures[110]. Disruption of sand transport by coastal protection structures, sewage pollution, beach nourishment, tourism, beach cleaning, bait collecting and fishing have previously been characterized as anthropogenic disturbances with negative consequences for the beach ecosystem[108]. Under the assumption that these human activities lead to similar ecological consequences and due to the difficulty of a distinct spatial separation of single elements, a common approach to evaluate human disturbances for beach ecosystems is the comparison between urbanized areas and remote, unaffected control sites[111–113]. However, it remains unclear whether various types of anthropogenic disturbances within urbanized areas (e.g. permanent settlements, infrastructure, tourist facilities, etc.) actually have similar impacts on the environment[114]. If not, then current conservation efforts might be improvable by developing strategies that are more specifically tailored to counteract the environmental degradation of the distinct human disturbance.

To investigate this question, the present study was conducted on small coral islands which were either (I) inhabited by the local population, (II) accommodating a tourist facility, or (III) completely uninhabited. This approach guaranteed a distinct spatial separation of two different anthropogenic disturbances and enabled a comparison to ecosystems with no permanent and direct human impact.

A terrestrial hermit crab community comprising two species (*C. rugosus* and *C. perlatus*) was chosen as an indicator to investigate human disturbances[115]. Terrestrial hermit crabs are a crucial component in beach ecosystems that link the marine and the terrestrial food web[56]. As adult terrestrial hermit crabs are restricted to the beaches, populations on small coral islands – like most beach-associated macrofauna – cannot avoid human stressors by migration[115]. Consequently, they can be considered representative of a large number of beach-associated taxa for the purpose of examining anthropogenic disturbances.

4.3   Results

4.3.1   Impact of Different Human Disturbances on the Abundance and Size of Hermit Crabs

_The studied organisms belonged to the only terrestrial genus of hermit crabs, *Coenobita*, and comprised *C. rugosus* and *C. perlatus*. Significant differences in the abundance and size of the investigated hermit crabs were observed between uninhabited, local and tourist islands (Fig. 6). Island type had a significant effect on the hermit crab abundance within the investigated plots (ANOVA: $N = 4$, $df = 2$, $F = 28.997$, $P < 0.001$). Significantly fewer hermit crabs were present in the plots on tourist islands than on uninhabited ($P > 0.001$) and local islands ($P < 0.001$). The abundance within the plots did not differ between uninhabited (16.25 ± 7.03 mean ± standard error) and local islands (17.87 ± 6.98; $P = 0.692$), although the availability of suitable habitats was significantly reduced on local islands, which might ultimately result in a reduced island population size on the local islands as well (see results section (c)). Furthermore, island type had a significant effect on the hermit crab size (ANOVA: $N = 4$, $df = 2$, $F = 5.764$, $P = 0.028$). On local islands, the investigated hermit crabs were significantly smaller compared to tourist islands ($P = 0.022$). No significant differences were observed between the size of hermit crabs on uninhabited (0.68 ± 0.01 cm) and on local islands (0.62 ± 0.02 cm; $P = 0.292$), nor between uninhabited and tourist islands (0.76 ± 0.04 cm; $P = 0.201$).

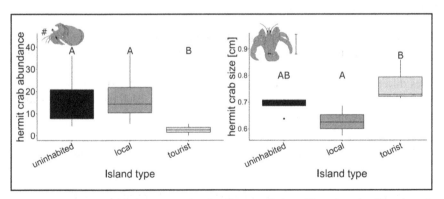

Figure 6: Anthropogenic impact on the abundance and size of hermit crabs. Hermit crab abundance (left) and hermit crab size (right) compared between uninhabited, local and tourist islands ($N = 4$). Significant differences between island types are indicated by different letters.

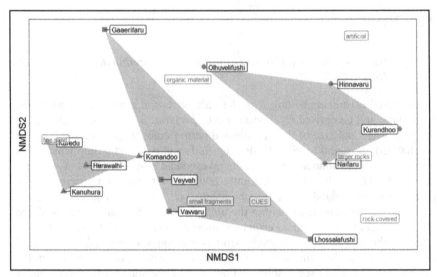

Figure 7: Distinctness of the three investigated island types. NMDS ordination of the investigated islands (blue squares and blue cluster area: uninhabited islands, red circles and red cluster area: local islands, green triangles and green cluster area: tourist islands) is based on the three resource and habitat parameters that influence hermit crab abundance and size (food, shell and habitat availability)). NMDS ordination thereby groups points, i.e. islands, with similar values closer together. Spatial proximity of a data point, i.e. an island, to one of the investigated parameters shows that the island is described by high values in the respective parameter.

To elicit potential reasons for the differences in hermit crab abundance and size between the three island types, food availability, beach habitat structure and empty shell resource were investigated using NMDS (Fig. 7). The three island types differed significantly in resource and habitat (PERMANOVA: $N = 4$, $df = 2$, $F = 4.770$, $P = 0.004$). For a more detailed analysis, each parameter was further investigated specifically:

4.3.2   Impact of Different Human Disturbances on the Food Resource of Hermit Crabs

Island type had no significant effect on the amount of organic material per m² on the beach (Kruskal-Wallis: $N = 4$, $df = 2$, $\chi^2 = 4.653$, $P = 0.097$), but calculated means suggest a non-significant tendency towards fewer organic material on tourist islands (1.14 ± 0.28 g), compared to uninhabited islands (4.63 ± 1.09 g) and local islands (2.85 ± 1.19 g).

4.3.3 Impact of Different Human Disturbances on the Beach Habitat Structure

The composition of the beach habitat (for categorization see methods section) varied significantly between the three island types (Fig. 8): the proportion of the fine sand beach habitat on the total island's circumference was significantly different between the three island types (Kruskal-Wallis: $N = 4$, $df = 2$, $\chi^2 = 7.565$, $P = 0.022$), with a significantly higher proportion of fine sand beach on tourist islands than on local islands ($P = 0.018$). Additionally, the proportion of artificial shoreline (Kruskal-Wallis: $N = 4$, $\chi^2 = 8.459$, $P = 0.014$) and vegetation-covered beach (Kruskal-Wallis: $N = 4$, $\chi^2 = 7.461$, $P = 0.024$) was significantly altered, with a significantly higher proportion of artificial shoreline on local islands than on uninhabited islands ($P = 0.013$) and significantly fewer vegetation-covered beach on tourist islands than on uninhabited islands ($P = 0.026$). No significant differences were observed in the proportion of "fine sand with small fragments" habitat (Kruskal-Wallis: $N = 4$, $\chi^2 = 0.115$, $P = 0.944$), "fine sand with larger rock" habitat (Kruskal-Wallis: $N = 4$, $\chi^2 = 4.832$, $P = 0.089$) and "predominantly rock-covered beach" habitat (Kruskal-Wallis: $N = 4$, $\chi^2 = 5.434$, $P = 0.066$). The adjacent shore composition did not differ significantly between the three island types (Kruskal-Wallis: Seagrass: $N = 4$, $\chi^2 = 0.927$, $P = 0.629$, Seagrass and Sand: $N = 4$, $\chi^2 = 1.457$, $P = 0.483$, Sand: $N = 4$, $\chi^2 = 0.731$, $P = 0.694$, Sand and Rock: $N = 4$, $\chi^2 = 2.457$, $P = 0.293$, Rock: $N = 4$, $\chi^2 = 4.352$, $P = 0.114$).

The investigated beach habitat types had a significant effect on the hermit crab abundance (crossed fixed-factor ANOVA island type x habitat type: $N = 4$, $df = 3$, $F = 5.969$, $P = 0.001$), but beach type and island type did not interact significantly ($N = 4$, $df = 5$, $F = 0.427$, $P = 0.827$). When considering the abundance of hermit crabs in only one of the four investigated beach habitat types, island type still had a significant effect on the hermit crab abundance: the abundance of hermit crabs in the "fine sand beach" habitat differed significantly between the three island types (Kruskal-Wallis: $N = 4$, $df = 2$, $\chi^2 = 15.920$, $P < 0.001$), with significantly fewer hermit crabs in the fine sand habitat of tourist islands than in that of uninhabited islands ($P < 0.001$) and of local islands ($P = 0.035$). Island type had also a significant effect on the abundance of hermit crabs in the "fine sand with small fragments beach" habitat (Kruskal-Wallis: $N = 4$, $df = 2$, $\chi^2 = 12.501$, $P = 0.001$) with significantly fewer hermit crabs in this habitat type on tourist islands than in uninhabited islands ($P = 0.007$) and local islands ($P = 0.007$).

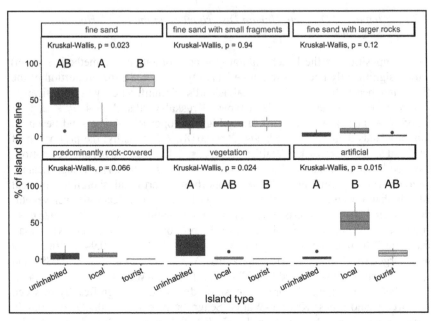

Figure 8: Beach habitat composition of the three island types. Proportions of each of the six categorized beach types on the three investigated island types ($N = 4$). Significant differences in the pairwise comparisons between island types are indicated by different letters.

4.3.4   Impact of Different Human Disturbances on the Shell Resource of Hermit Crabs

Island type had a significant effect on the overall abundance of empty shells (Kruskal-Wallis: $N = 4$, $df = 2$, $\chi^2 = 7.130$, $P = 0.028$) and on the crab-per-utilizable-empty-shell (CUES)-ratio (Kruskal-Wallis: $N = 4$, $df = 2$, $\chi^2 = 7.730$, $P = 0.020$). This CUES-ratio can be understood as a measure for the intensity of competition over the shell resource. Higher values of this ratio indicate a more severe competition, while values closer to 1 indicate that for each hermit crab a potential utilizable empty shell is readily available. The CUES-ratio was signifi-cantly smaller on tourist islands than on uninhabited islands ($P = 0.024$). On uninhabited islands, on average 10 hermit crabs competed over one shell, while on local islands only 6 hermit crabs competed over one shell. For each hermit crab on a tourist island existed on average one utilizable empty shell. Island type had a significant effect on the abundance of non-utilizable empty shells (Krus-kal-Wallis: $N = 4$, $df = 2$, $\chi^2 = 6.545$, $P = 0.037$): significantly more non-utilizable empty shells were found on local islands than on uninhabited islands $P = 0.046$) and on tourist islands ($P = 0.046$), while the number of non-utilizable empty shells did not differ statistically between uninhabited and tourist islands ($P = 0.922$). To investigate the reasons for the hermit crab size differences, the shell parameter that most strongly determines hermit crab size, i.e. the aperture area of the shell (Spearman: $R^2 = 0.861$, $P < 0.001$), was analysed. The aperture area of utilized shells did not differ significantly between the three island types (Kruskal-Wallis: $N = 4$, $df = 2$, $\chi^2 = 5.303$, $P = 0.070$). The aperture area of utilizable empty shells did not differ between the three island types (Kruskal-Wallis: $N = 4$, $df = 2$, $\chi^2 = 3.803$, $P = 0.149$).

4.4   Discussion

Numerous studies have demonstrated that coastal ecosystems are substantially altered or degraded in urbanized areas[111-113]. Due to spatial proximity, different anthropogenic disturbances impact beach ecosystems simultaneously in those areas. It is therefore difficult to disentangle the environmental impacts of different disturbances and investigate with certainty whether ecosystems respond differently to different disturbances[114]. We investigated this issue by studying small coral islands, where different anthropogenic disturbances are spatially separated. The results from our novel study approach show that these disturbances are having clear but distinct impacts on the investigated terrestrial hermit crabs. These findings, based on our study approach, should be transferable to a large number of beach-dwelling taxa, as food and habitat availability generally limits species distribution and population size [101,116].

   On tourist islands, hermit crabs were significantly less abundant and significantly larger than on local islands. Compared to the uninhabited reference system, the abundance was negatively impacted on tourist islands, but did not differ compared to local islands. However, the overall population size on local islands should be considered reduced, as the availability of suitable habitats has been reduced by harbours and coastal protection structures. Therefore, different elements of urbanized areas, i.e. permanent settling or tourism, can have distinct environmental impacts on beach ecosystems.

   Food, habitat availability and empty shell abundance are limiting resources for hermit crabs and might offer reasons for the observed differences between the two different land uses[38,101]. The tendency towards less organic material on tourist islands (1.14 ± 0.31 g/m²) compared to local islands (4.26 ± 3.43 g/m²) and the uninhabited reference (4.88 ± 1.84 g/m²) could be explained by beach grooming measurements, which were performed on all four studied resort islands up to four times per day (personal communication). Beach grooming is a common practice around tourist facilities and aims to remove washed-up organic material and debris from the beaches[114]. It causes a reduced food availability for the affected beach fauna, which can result in decreased population densities[117]. In concordance, on average only three hermit crabs per plot were found on the groomed beaches of the tourist islands, compared to 16 hermit crabs on average on uninhabited islands. The beach fauna on the tourist islands might also experience a higher mortality from the cleaning process, either when getting accidentally removed together with the algal material (personal observation) or when being mechanically crushed in the cleaning process, as already demonstrated for ghost crabs[118]. Hence, we hypothesize that beach cleaning is one reason for the significantly decreased abundance on islands with tourist facilities. As beach cleaning was not performed on local islands, hermit crab abundance in suitable habitats remained unaffected (average 18 hermit crabs per plot), although beaches are also used by the local population for recreational activities.

Apart from the overall availability of organic material, the beach habitat structure needs to be considered when investigating the population structure of the beach fauna[119]: compared to the structurally more complex beach habitat types, the fine sand beaches had a significantly reduced hermit crab abundance on all three island types. On tourist islands, this fine sand beach habitat accounted for $75 \pm 12\%$ of the total circumference. However, the higher proportion of the more sparsely inhabited fine sand beach cannot be held solely responsible for the reduced hermit crab abundance on tourist islands. Less than one hermit crab per plot was collected in the fine sand beach habitat of tourist islands, while on average eleven hermit crabs were present in the fine sand beach habitat on uninhabited islands. Therefore, disturbances associated with tourist facilities are probably responsible for the reduced abundance on the fine sand beaches of tourist islands. Beach nourishment, a technique where sand gets extracted from the adjacent benthic zone and deposited on the existing shoreline to extend the sandy beaches desired by tourists, is often performed to an extent where the whole natural beach shoreline becomes artificially altered to unvegetated sandy beaches[120]. This measurement can reduce the population size of the whole beach fauna[115,121,122] – especially when the beach-associated vegetation is completely removed, many beach taxa can become completely absent[115,120]. Therefore, we hypothesize that the removal of beach-associated vegetation, together with the removal of organic material caused by beach grooming and nourishment, are the main drivers for the reduced hermit crab abundance on the islands with tourist facilities.

The shoreline of local islands was differently altered and affected than that of tourist islands: the shoreline of local islands was $53 \pm 21\%$ artificially obstructed in form of concrete walls, either for harbour sites or to stabilize reclaimed land. Hence, on average only about half of the local islands shoreline formed a soft-bottom beach habitat suitable for beach-associated organisms[21]. Although the abundance in the investigated plots on local islands were similar to those on uninhabited islands, local islands as a whole, with their extensive artificial shorelines, must be considered as degraded coastal ecosystems with reduced and fragmented beach habitats[123]. In conclusion, this suggests that the total hermit crab population size of a complete local island is on average 50% smaller than the overall population size of uninhabited islands, as the constructions on local islands caused the shoreline to become widely uninhabitable for these organisms[5]. However, the hermit crab abundance within suitable beach habitats did not differ between uninhabited and local islands. This demonstrates that beach-dwelling organisms can occur in densely populated areas in the same high abundance as they do on uninhabited islands, as long as the beach habitat itself remains intact and not altered by human activities.

Besides food availability and habitat structure, shell availability is the most limiting resource for hermit crabs, as they are dependent on the input of empty gastropod shells from the adjacent coastal waters[15]. Therefore, analysing patterns in the shell resource might offer further explanations for the observed differences between the different island types.

The number of non-utilizable empty shells, like cones or cowries, can be considered as a proxy for the overall shell input of an island as these shells accumulate on the beaches without getting removed or utilized by hermit crabs[124]. The number of non-utilizable empty shells did not differ between uninhabited and tourist islands, suggesting that the overall input of the shell resource was similar on both island types. Taken together with the significantly reduced CUES-ratio on tourist islands (on average, one utilizable empty shell per hermit crab was available), neither a diminished shell input, nor high competition over the shell resource, are responsible for the significantly decreased population densities on tourist islands. A sufficient number of empty shells can result in a strong growth of a hermit crab population in a natural system[15]. This suggests that, based on the availability of the shell resource, populations on the tourist islands would have the potential to further grow, but are probably limited due to beach grooming or removal of vegetation.

On local islands however, the number of non-utilizable empty shells was on average four times higher than on uninhabited islands. Harvesting of gastropods for consumption has been shown to provide a surplus of empty gastropod shells for hermit crab populations and might be responsible for the overall increase in shells on local island[125]. Furthermore, an overall higher gastropod population density in the adjacent coastal waters might be an additional reason for the increased empty shell abundance. This might stem from a greater food supply resulting from wastewater release[126]. This effect only occurred on the local islands, as sewage and other municipal waste is released mostly untreated into the coastal water, while tourist resorts collect the effluents in septic tanks, thereby minimizing nutrient enrichment of the adjacent waters[127].

The higher abundance of empty gastropod shells on local island beaches is beneficial for the hermit crab populations, as the limiting resource becomes largely available[16]. This is also shown by a decreased CUES-ratio on local islands, suggesting a reduced competition over the shell resource compared to tourist islands and the uninhabited reference. This could explain at least partially why the hermit crab abundance within the investigated plots remained unaffected on the local islands in the present study.

Although the abundance within the investigated plots was not affected negatively, the mean body size on local islands was decreased compared to tourist islands. The body size of a hermit crab correlated with the aperture area of its utilized shell. Therefore, analysing the aperture area of the utilizable empty shells might provide an explanation for the reduced body size on local islands, as the size of the aperture limits growth[4]. However, the aperture areas of both the utilized shells and the utilizable empty shells did not differ significantly between the three island types. This suggests that a lack of larger empty shells is not the main driver for the reduced body size in hermit crabs on local islands, as enough large-sized shells were available, potentially allowing the hermit crabs on the local islands to further grow. Therefore, we hypothesize that human activities on the local islands are responsible for the reduced body size: beach-dwelling decapod crustaceans, like *C. perlatus*, are widely used as fishing bait by the local fishermen[128]. They may select for bigger specimen, as they are easier to find and more suitable as fishing bait[129]. A size-selective harvesting could result in smaller body sizes on local islands, compared to uninhabited and tourist islands, where harvesting is absent[51]. A comparable human-driven size selection is already known in commercial gastropod and fish species, where intensive harvesting and fishing resulted in a shift towards smaller body size due to overexploitation of the larger-sized specimen[130,131]. In comparison, hermit crabs were significantly larger on tourist islands. This can be linked to the reduced abundance on these islands, as a smaller population size decreases intraspecific competition, which ultimately can enable organisms to grow larger[56].

Our study reveals that two elements of urbanized areas have different environmental impacts. Abundance was negatively impacted on tourist islands, whereas body size was negatively impacted on local islands. Although the abundance within the investigated plots was unaffected on local islands, it is negatively impacted on a larger scale, as about half of the shoreline consists of concrete walls for harbour sites and coastal protection and is therefore uninhabitable for all beach-dwelling organisms.

Here, it is demonstrated that the environment is not always impacted identically by the different elements of an urbanized area, but rather that the type of anthropogenic disturbance is decisive for the ecological consequence. At the same time, organisms can maintain the same population size in densely populated areas as in uninhabited ecosystems, as long as certain habitat characteristics remain unaffected. Our novel approach using small islands thereby ensured that the observed environmental impacts are attributable to only one element of an urbanized area, namely tourism or permanent settlement.

The implications of this study are beneficial for environmental protection measures, as it demonstrates the importance of disentangling various types of disturbance that stem from urbanized areas and to consider each element specifically when developing management strategies for conservation[132]. In practical terms this could mean that the prime measurement for tourist facilities is to reduce beach grooming and leave seagrass and other allochthonous material as a food resource for the beach fauna. The prime measurement for permanently colonized land on the other hand would be to minimize the obstruction of the shoreline by concrete structures and implement some regulations that leave parts of the shoreline as natural sandy beaches. These two proposed management implications to counteract two different forms of land use underline how important it is to disentangle anthropogenic disturbances. A greater understanding of how specific human actions lead to certain environmental responses, will enable us to better curtail these stressors and counteract the global loss of biodiversity and ecosystems[133].

4.5   Methods

The research was conducted under the permission of the Ministry of Fisheries and Agriculture (Male', Maldives), permit number: (OTHR)30-D/INDIV/2017/ 122 and in accordance with the given guidelines and regulations.

Sampling was carried out on 12 small coral islands, all located within the Lhaviyani (Faadhippolhu) Atoll, Republic of Maldives. The islands were assigned into three categories: (I) islands that were inhabited solely by the local Maldivian population (local islands), (II) islands with a tourist resort (tourist island) and (III) islands with no permanent direct human disturbance (uninhabited islands). Note that Vavvaru island is strictly speaking not a completely uninhabited island but was a former marine biology field station (Korallionlab). However, the station has closed and during its active time only inhabited three to five staff members and occasionally guest researchers. Sampling of the island's beaches was carried out from 03/02/2017 to 10/03/2017, always within 2 hours before low tide until low tide. The whole island's beachline and the adjacent shore were mapped with GPS (eTrex Vista® Cx, Garmin Ltd., Schaffhausen, Switzerland) by assigning it in the following habitat categories: artificial, vegetation-covered (i.e. inaccessible beach, covered fully by shrub vegetation), predominantly rock-covered beach, fine sand with larger rocks, fine sand with small fragments and fine sand beach for the beachline and seagrass, seagrass & sand, sand, sand & rock, rock for the adjacent shore. The percentage of each habitat on the total circumference of each island was calculated.

Each beach was sampled in the abovementioned beach habitat types, distributed randomly over the natural shoreline of the island. The vegetation-covered beach habitat and artificial shorelines were excluded from the sampling due to their inaccessibility. To minimize a biased selection of the sampled part of the beach, the location of the plot was chosen from a distance of minimum 15 m, so that the present hermit crabs could not have been seen in advance. The sampling plots were chosen to guarantee that each present beach habitat type was sampled at least once. Additionally, the two dominant habitat types (i.e. highest percentage of the islands circumference) of every island were sampled in a second plot. When one habitat type was not present on an island or covered less than 10 m in length (i.e. the plot size), an additional plot within the dominant habitat type was sampled, resulting in a total of six plots per island.

Each plot was 10 m long and 2 m wide, measured landwards from the present drift line using a folding rule and a measuring tape. The position of every plot was documented using GPS. All hermit crabs and all empty shells within the plot were counted, collected and stored in a plastic bucket for further analysis.

To assess the amount of potential food, the organic debris in four 0.5 m x 0.5 m sub-plots (resulting in 1 m² per plot in total) within each plot was collected using forceps and stored in a plastic bag. The four sub-plots were positioned at equal distances in a diagonal manner within the plot (0 m, 3.3 m, 6.6 m and 10 m along the plot length and at distances of 1.5 m, 1.0 m, 0.5 m and 0 m from the drift line). The wet weight of the organic material per plot was measured using a fine scale (TS-300 300g x 0.01 g, G&G GmbH, Neuss, Germany).

Hermit crabs were removed from their shell by carefully heating the apex of the shell above an open flame. This is a standard procedure to remove hermit crabs from their shells and leaves the animal without injuries[47,64]. Hermit crabs were photographed on millimetre paper (Nikon D5000 mounted with Nikon AF-S Nikkor 18-105 mm, 1:3.5-5.6, Nikon Corp., Tokyo, Japan.).

All shells (utilized and empty) were photographed on millimetre paper and identified using morphological identification keys[65,66,68,69]. All empty shells were assigned in two categories: (I) empty shells belonging to a gastropod species that was found to be utilized by a hermit crab and therefore considered being in general utilizable, and (II) empty shells belonging to a gastropod species, which was never found to be utilized by a hermit crab (mainly cone or cowrie shells) and therefore considered to be generally not utilizable by the investigated hermit crab species. Non-utilizable empty shells, like cowrie or cone shells, accumulate on the beaches without being ever utilized or transferred over longer distances by hermit crabs or any other beach inhabitant[13,124] and can therefore be used as a proxy for the overall shell input on the beaches

After this procedure, the hermit crabs were transferred into a plastic bucket together with their removed shell and left to recover before being transferred back to their original beach habitat.

The size of the hermit crabs and their corresponding shell was determined using ImageJ 1.49b (Rasband, W.S., ImageJ, U. S. National Institutes of Health, Bethesda, Maryland, USA, http://imagej.nih.gov/ij/, 1997-2015) by measuring the carapace length of the hermit crab, and the length and width of the aperture area of each shell.

The statistical analysis was carried out using R 3.5.1, extended with the "vegan" package for multivariate ecological analysis[72]. Prior to statistical analysis, abundance data was Tukey-transformed (lambda = 0.375) to meet the assumptions of normality and variance homogeneity. Where assumptions for parametric testing were violated, non-parametric Kruskal-Wallis tests were conducted. To test for differences in hermit crab abundance between the three island types (uninhabited, local, tourist islands) and account for the different habitat types on each island, univariate ANOVA with crossed fixed factors (island type x habitat type) was performed and pairwise comparisons were calculated using TukeyHSD post-hoc tests ($N = 4$). The Influence of human land use on hermit crab size was analysed by calculating the mean body size for each island and statistically compare it between the three island types ($N = 4$) using ANOVA and TukeyHSD post-hoc tests. To investigate, how the two different forms of human land use influence the underlying resources of hermit crabs, a non-metric multi-dimensional scaling (NMDS) was performed. First, the parameters "empty shell abundance", "organic material" and the proportion of the four different beach habitat types were rescaled between 0 and 1 for Bray-Curtis dissimilarity matrix calculation. Then, NMDS ordination was calculated using $k = 2$ dimensions. To test for differences in resource availability between the three island types based on the NMDS, a PERMANOVA was calculated (Bray-Curtis, 4999 permutations). Additionally, Kruskal-Wallis tests with Dunn post-hoc tests and Bonferroni corrections were performed to compare the underlying resources (i.e. organic material [g/m^2], empty shell abundance, and proportion of each beach habitat type) separately between the three island types ($N = 4$). The abundance of hermit crabs within the "fine sand with larger rocks"- and the "predominantly rock-covered"-beach habitat were not compared individually between the three island types, as the "fine sand with larger rocks"-habitat occurred only on 50% of all investigated islands and the predominantly rock-covered beach was overall absent on tourist islands. To further investigate reasons for the differences in hermit crab size between the three island types, the shell parameter that correlated best with hermit crab size was identified using Spearman rank correlation test. The aperture area of the shell showed a high correlation with hermit crab body size ($R^2 = 0.861$, $P < 0.001$) and was subsequently compared for utilized and utilizable empty shells between the three island types using Kruskal-Wallis tests.

5 Synopsis

The terrestrial hermit crabs of the genus *Coenobita* take up an important role for the beach environment as cleaning agents and as a link between the marine and terrestrial food web[21–23]. As comparably little is known about the biology of this significant taxa, the aim of this master thesis was to better understand the mechanisms and factors that influence the populations of these tropical crustaceans. Populations of the two sympatric species *C. rugosus* and *C. perlatus* were thereby shown to be influenced by biotic, abiotic and anthropogenic factors in their behaviour, abundance and size structure:

Concluding from the shell selection experiments, *C. rugosus* and *C. perlatus* had no observable interspecific competition over certain shell species, but instead seem to have evolved two distinct shell selection behaviours, where the preferred shells of one species are the neglected shells of the second. This allows the two beach-associated hermit crabs to coexist, rather than compete over the strictly limited shell resource in their natural habitat. Factors affecting the relative abundances of *C. rugosus* and *C. perlatus* are therefore not caused by the shell availability and interspecific competition, but might arise out of other, behavioural differences, like the tendency to disperse further landwards or the avoidance of mangrove forest-covered coastlines[57,88].

Abiotic factors that were identified to strongly influence the abundance, were the tides and daytime. For a beach-inhabiting organism, the risk to desiccate or to be mechanically disrupt by wave action is omnipresent[84]. To minimize these risks, *Coenobita* hermit crabs have evolved behavioural mechanism, due to which they become virtually inactive during the hottest time of day, i.e. midday. The time of their highest activity however lays in the hour before low tide until low tide, where the risk of displacement by incoming waves is small, while food availability is high[83,134]. To further optimize their scavenging behaviour, hermit crabs gather in the structurally more complex beach habitats, as food availability in these areas is increased[103].

Besides these natural, abiotic factors that influence hermit crabs, human land use has a major impact on their populations. Especially on the heavily used tourist islands, the hermit crab abundances were less than one-fourth of those found on uninhabited islands. The local island use on the other hand does not directly impact the abundance of the hermit crab populations, but strongly influences the size structure of the hermit crab population, as the largest specimen are virtually absent, probably as a result of being used as bait for fishery[14].

Hermit crab populations on small coral island are highly variable in space and time. They are strongly influenced by their abiotic environment, by the pres-

© The Editor(s) (if applicable) and The Author(s), under exclusive license to
Springer Fachmedien Wiesbaden GmbH, part of Springer Nature 2020
S. Steibl, *Terrestrial Hermit Crab Populations in the Maldives*, BestMasters,
https://doi.org/10.1007/978-3-658-29541-7_5

ence of other hermit crab species and by the presence and type of human land use. While coenobitid hermit crabs evolved behavioural adaptation to withstand abiotic stressors, like heat, and competition over the limiting shell resource, they were unable to adapt to the rapid expansion and intensive exploitation of the island ecosystems by humans. Unlike on the continental shorelines, small islands do not have the possibility for their fauna to avoid human activities by moving away from the anthropogenic stressor[135]. Therefore, when human exploitation of small islands negatively impacts the ecosystem, it leads to much higher environmental impacts than on the mainland, as the inhabiting animals lose their habitat without the possibility to move into adjacent, more well-preserved areas.

Terrestrial hermit crabs, could therefore also act as a good proxy for a rapid evaluation of human land use on small island ecosystems[136,137]. Possible impacts on the beaches can be analysed by investigating the abundance, size and diversity of hermit crab populations and combined with the possible impacts in the adjacent waters, by also analysing the abundance, size and diversity of their utilized gastropod shells. To establish hermit crabs as an evaluation tool for island ecosystem health, more detailed studies will be necessary that investigate thoroughly, how human activities directly influence hermit crabs and their utilized gastropod shells. Nevertheless, it is also essential to further analyse the natural biotic and abiotic factors that influence hermit crab populations and gastropod communities to be able to distinguish between natural variations in their distributions and those caused by anthropogenic stressors.

Being able to rapidly and easily assess the status quo of island ecosystems is desirable for conservation efforts and managements. Using hermit crabs for ecosystem evaluations is a novel approach in conservation biology that combines the advantageous of being a fast method and investigating organisms that are easy to collect and identify. By also considering the gastropod shells, conclusions about the status quo of the ecosystem in the adjacent waters can be drawn without expensive and time-consuming scuba diving programs. Combining a terrestrial and a marine taxon could therefore be the key advantage of hermit crabs in becoming an ecosystem indicator species for conservation work in the future.

References

1. Williams, J. D. & McDermott, J. J. *Hermit crab biocoenoses: a worldwide review of the diversity and natural history of hermit crab associates*. Journal of Experimental Marine Biology and Ecology **305**, (2004).

2. Peura, J. F., Lovvorn, J. R., North, C. A. & Kolts, J. M. Hermit crab population structure and association with gastropod shells in the northern Bering Sea. *J. Exp. Mar. Bio. Ecol.* **449**, 10–16 (2013).

3. McLaughlin, P. A. Hermit crabs: are they really polyphyletic? *J. Crustac. Biol.* **3**, 608–621 (1983).

4. Hazlett, B. A. The behavioral ecology of hermit crabs. *Annu. Rev. Ecol. Syst.* **12**, 1–22 (1981).

5. Reese, E. S. Behavioral adaptations of intertidal hermit crabs. *Am. Sci.* **9**, 343–355 (1969).

6. Blackstone, N. W. Specific growth rates of parts in a hermit crab: a reductionist approach to the study of allometry. *J. Zool.* **211**, 531–545 (1987).

7. Brodie, R. J. Ontogeny of shell-related behaviors and transition to land in the terrestrial hermit crab *Coenobita compressus* H. Milne Edwards. *J. Exp. Mar. Bio. Ecol.* **241**, 67–80 (1999).

8. Bertness, M. D. The influence of shell-type on hermit crab grwoth rate and clutch size (Decapoda, Anomura). *Crustaceana* **40**, 197–205 (1981).

9. Conover, M. R. The importance of various shell characteristics to the shell-selection behavior of hermit crabs. *J. Exp. Mar. Biol. Ecol* **32**, 131–142 (1978).

10. Fotheringham, N. Population consequences of shell utilization by hermit crabs. *Ecology* **57**, 570–578 (1976).

11. Herreid, C. F. & Full, R. J. Energetics of hermit crabs during locomotion: the cost of carrying a shell. *J. Exp. Mar. Bio. Ecol.* **120**, 297–308 (1986).

12. Osorno, J.-L., Fernández-Casillas, L. & Rodríguez-Juárez, C. Are hermit crabs looking for light and large shells?: evidence from natural and field induced shell exchanges. *J. Exp. Mar. Bio. Ecol.* **222**, 163–173 (1998).

13. Völker, L. Zur Gehäusewahl des Land-Einsiedlerkrebses *Coenobita scaevola* Forskal vom Roten Meer. *J. Exp. Mar. Bio. Ecol.* **1**, 168–190 (1967).

14. de Wilde, P. A. W. J. On the ecology of *Coenobita clypeatus* in Curacao. *Stud. Fauna Curaçao other Caribb. Islands* **44**, 1–138 (1973).

15. Vance, R. R. Competition and mechanism of coexistence in three sympatric species of intertidal hermit crabs. *Ecology* **53**, 1062–1074 (1972).

16. Scully, E. P. The effects of gastropod shell availability and habitat characteristics on shell utilization by the intertidal hermit crab *Pagurus longicarpus* Say. *J. Exp. Mar. Biol. Ecol* **37**, 139–152 (1979).

17. Grant, W. C. & Ulmer, K. M. Shell selection and aggressive behavior in two sympatric species of hermit crabs. *Biol. Bull.* **146**, 32–43 (1974).

18. Angel, J. E. Effects of shell fit on the biology of the hermit crab Pagurus longicarpus (Say). *J. Exp. Mar. Bio. Ecol.* **243**, 169–184 (2000).

19. McMahon, B. R. & Burggren, W. W. Respiration and adaptation to the terrestrial habitat in the land hermit crab *Coenobita clypeatus*. *J. exp. Biol* **79**, 265–281 (1979).

20. Greenaway, P. Terrestrial adaptations in the Anomura (Crustacea: Decapoda). *Mem. Museum Victoria* **60**, 13–26 (2003).

21. Page, H. M. & Willason, S. W. Distribution patterns of terrestrial hermit crabs at Enewetak Atoll, Marshall Islands. *Pacific Sci.* **36**, 107–117 (1982).

22. Ball, E. E. Observations on the biology of the hermit crab *Coenobita compressus* H. Milne Edwards (Decapoda; Anomura) on the west coast of the Americas. *Rev. Biol. Trop.* **20**, 265–273 (1972).

23. Page, H. M. & Willason, S. W. Feeding activity patterns and carrion removal by terrestrial hermit crabs at Enewetak Attol, Marshall Islands. *Pacific Sci.* **37**, 151–155 (1983).

24. Burggren, W. W. & McMahon, B. R. *Biology of the land crabs*. (Cambridge University Press, 1988). doi:10.1016/0169-5347(89)90078-5

25. Steibl, S. & Laforsch, C. Shell resource partitioning as a mechanism of coexistence in two co-occurring terrestrial hermit crab species. *BMC Ecol.* **20**, (2020).

26. Steibl, S. & Laforsch, C. Daytime activity and habitat preferences of two sympatric hermit crab species (Decapoda: Anomura: *Coenobita*). *Estuar. Coast. Shelf Sci.* **231**, 106482 (2019).

27. Steibl, S. & Laforsch, C. Disentangling the environmental impact of different human disturbances: a case study on islands. *Sci. Rep.* 13712 (2019). doi:10.1038/s41598-019-49555-6

28. Barnes, D. K. A. Local, regional and global patterns of resource use in ecology: hermit crabs and gastropod shells as an example. *Mar. Ecol. Prog. Ser.* **246**, 211–223 (2003).

29. Birch, L. C. The meanings of competition. *Am. Nat.* **91**, 5–18 (1957).

30. Klomp, H. The concepts 'similar ecology' and 'competition' in animal ecology. *Arch. Neerl. Zool.* **14**, 90–102 (1961).

31. Abrams, P. A. Shell selection and utilization in a terrestrial hermit crab, Coenobita compressus (H. Milne Edwards). *Oecologia* **34**, 239–253 (1978).

32. Roughgarden, J. Resource partitioning among competing species - a coevolutionary approach. *Theor. Popul. Biol.* **9**, 388–424 (1976).

33. Schoener, T. W. Resource partitioning in ecological communities. *Science (80-.)*. **185**, 27–39 (1974).

34. Hardin, G. The competitive exclusion principle. *Science (80-.)*. **131**, 1292–1297 (1960).

35. Gherardi, F. & Nardone, F. The question of coexistence in hermit crabs: population ecology of a tropical intertidal assemblage. *Crustaceana* **70**, 608–629 (1997).

36. MacArthur, R. H. & Levins, R. Competition, habitat selection, and character displacement in a patchy environment. *PNAS* **51**, 1207–1210 (1964).

37. Abrams, P. A. Resource partitioning and interspecific competition in a tropical hermit crab community. *Oecologia* **46**, 365–379 (1980).

38. Fotheringham, N. Hermit crab shells as a limiting resource (Decapoda, Paguridea). *Crustaceana* **31**, 193–199 (1976).

39. Hazlett, B. A. Interspecific shell fighting in three sympatric species of hermit crabs in Hawaii. *Pacific Sci.* **24**, 472–482 (1970).

40. Kavita, J. Spatial and temporal variations in population dynamics of few key rocky intertidal macrofauna at tourism influenced intertidal shorelines. (Saurashtra University, 2010).

41. Borjesson, D. L. & Szelistowski, W. A. Shell selection, utilization and predation in the hermit crab *Clibanarius panamensis* stimpson in a tropical mangrove estuary. *J. Exp. Mar. Bio. Ecol.* **133**, 213–228 (1989).

42. Vance, R. R. The role of shell adequacy in behavioral interactions involving hermit crabs. *Ecology* **53**, 1075–1083 (1972).

43. Bach, C., Hazlett, B. A. & Rittschof, D. Effects of interspecific competition on fitness of the hermit crab *Clibanarius tricolor*. *Ecology* **57**, 579–586 (1976).

44. Childress, J. R. Behavioral ecology and fitness theory in a tropical hermit crab. *Ecology* **53**, 960–964 (1972).

45. Bertness, M. D. Shell preference and utilization patterns in littoral hermit crabs of the bay of Panama. *J. Exp. Mar. Bio. Ecol.* **48**, 1–16 (1980).

46. Gherardi, F. & McLaughlin, P. A. Shallow-water hermit crabs (Crustacea: Decapoda: Anomura: Paguridea) from Mauritius and Rodrigues Islands, with the description of a new species of Calcinus. *Raffles Bull. Zool.* **42**, 613–656 (1994).

47. Reddy, T. & Biseswar, R. Patterns of shell utilization in two sympatric species of hermit crabs from the Natal Coast (Decapoda, Anomura, Diogenidae). *Crustaceana* **65**, 13–24 (1993).

48. Blackstone, N. W. The effects of shell size and shape on growth and form in the hermit crab *Pagurus longicarpus*. *Biol. Bull.* **168**, 75–90 (1985).

49. Wilber Jr., T. P. & Herrnkind, W. Rate of new shell acquisition by hermit crabs in a salt marsh habitat. *J. Crustac. Biol.* **2**, 588–592 (1982).

50. Mitchell, K. A. Shell selection in the hermit crab *Pagurus bernhardus*. *Mar. Biol.* **35**, 335–343 (1976).

51. Hsu, C.-H. & Soong, K. Mechanisms causing size differences of the land hermit crab *Coenobita rugosus* among eco-islands in Southern Taiwan. *PLoS One* **12**, e0174319 (2017).

52. Nigro, K. M. *et al.* Stable isotope analysis as an early monitoring tool for community-scale effects of rat eradication. *Restor. Ecol.* 1–11 (2017). doi:10.1111/rec.12511

53. Sallam, W. S., Mantelatto, F. L. M. & Hanafy, M. H. Shell utilization by the land hermit crab Coenobita scaevola (Anomura, Coenobitidae) from Wadi El-Gemal, Red Sea. *Belgian J. Zool.* **138**, 13–19 (2008).

54. Willason, S. W. & Page, H. M. Patterns of shell resource utilization by terrestrial hermit crabs at Enewetak Atoll, Marhsall Islands. *Pacific Sci.* **37**, 157–164 (1983).

55. Kadmon, R. & Allouche, O. Integrating the effects of area, isolation, and habitat heterogeneity on species diversity: a unification of island biogeography and niche theory. *Am. Nat.* **170**, 443–454 (2007).

56. Morrison, L. W. & Spiller, D. A. Land hermit crab (*Coenobita clypeatus*) densities and patterns of gastropod shell use on small Bahamian islands. *J. Biogeogr.* **33**, 314–322 (2006).

57. Barnes, D. K. A. Hermit crabs, humans and Mozambique mangroves. *Afr. J. Ecol.* **39**, 241–248 (2001).

58. Gross, W. J. Water balance in anomuran land crabs on a dry atoll. *Biol. Bull.* **126**, 54–68 (1964).

59. Hsu, C.-H., Otte, M. L., Liu, C.-C., Chou, J.-Y. & Fang, W.-T. What are the sympatric mechanisms for three species of terrestrial hermit crab (Coenobita rugosus, C. brevimanus, and C. cavipes) in coastal forests? *PLoS One* **13**, e0207640 (2018).

60. Vannini, M. Researches on the coast of Somalia. The shore and the dune of Sar Uanle 10. Sandy beach decapods. *Monit. Zool. Ital.* **8**, 255–286 (1976).

61. Barnes, D. K. A. Ecology of tropical hermit crabs at Quirimba Island, Mozambique: shell characteristics and utilisation. *Mar. Ecol. Prog. Ser.* **183**, 241–251 (1999).

62. Lively, C. M. A graphical model for shell-species selection by hermit crabs. *Ecology* **69**, 1233–1238 (1988).

63. Bertness, M. D. Conflicting advantages in resource utilization: the hermit crab housing dilemma. *Am. Nat.* **118**, 432–437 (1981).

64. Bertness, M. D. Predation, physical stress, and the organization of a tropical rocky intertidal hermit crab community. *Ecology* **62**, 411–425 (1981).

65. Abbott, R. T. & Dance, S. P. *Compendium of Seashells.* (E.P. Dutton Inc., 1983).

66. Bosch, D. T., Dance, S. P., Moolenbeek, R. G. & Oliver, P. G. *Seashells of Eastern Arabia.* (Motivate Publishing, 1995).

67. Hogarth, P. J., Gherardi, F. & McLaughlin, P. A. Hermit crabs (Crustacea Decapoda Anomura) of the Maldives with the description of a new species of Catapagurus A. Milne Edwards 1880. *Trop. Zool.* **11**, 149–175 (1998).

68. Okutani, T. *Marine Mollusks in Japan.* (Tokai University Press, 2000).

69. Steger, J., Jambura, P. L., Mähnert, B. & Zuschin, M. Diversity, size frequency distribution and trophic structure of the macromollusc fauna of Vavvaru Island (Faadhippolhu Atoll, northern Maldives). *Ann. des naturhistorischen Museums Wien* **119**, 17–54 (2017).

70. Team, R. R: A language and environment for statistical computing. (2013).

71. Fisher, R. A. The logic of inductive inference. *J. R. Stat. Soc.* **98**, 39–82 (1935).

72. Oksanen, J. Multivariate analysis of ecological communities in R: vegan tutorial. *R Doc.* 43 (2015). doi:10.1016/0169-5347(88)90124-3

73. Latreille, P. A. *Les Crustacés, les Arachnides et les Insectes, distribués en familles naturelles, ouvrage formant les tomes 4 et 5 de celui de M. le Baron Cuvier sur le Règne animal (deuxième édition).* (1829).

74. Defeo, O. & McLachlan, A. Patterns, processes and regulatory mechanisms in sandy beach macrofauna: a multi-scale analysis. *Mar. Ecol. Prog. Ser.* **295**, 1–20 (2005).

75. Short, A. D. The role of wave height, period, slope, tide range and embaymentisation in beach classifications: a review. *Rev. Chil. Hist. Nat.* **69**, 589–604 (1996).

76. McLachlan, A. Dissipative beaches and macrofauna communities on exposed intertidal sands. *J. Coast. Res.* **6**, 57–71 (1990).

77. Noy-Meir, I. Structure and function of desert ecosystems. *Isr. J. Bot.* **28**, 1–19 (1980).

78. Hendrickx, M. E. Habitats and biodiversity of decapod crustaceans in the SE Gulf of California, México. *Rev. Biol. Trop.* **44**, 603–617 (1996).

79. Leite, F. P. P., Turra, A. & Gandolfi, S. M. Hermit crabs (Crustacea: Decapoda: Anomura), gastropod shells and environmental structure: their relationship in southeastern Brazil. *J. Nat. Hist.* **32**, 1599–1608 (1998).

80. McLachlan, A. & Dorvlo, A. Global patterns in sandy beach macrobenthic communities. *J. Coast. Res.* **21**, 674–687 (2005).

81. McLachlan, A., Jaramillo, E., Donn, T. E. & Wessels, F. Sandy beach macrofauna communities and their control by the physical environment: a geographical comparison. *J. Coast. Res.* **Special Is**, 27–38 (1993).

82. Felicita, S. Behaviour of mobile macrofauna is a key factor in beach ecology as response to rapid environmental changes. *Estuar. Coast. Shelf Sci.* **150**, 1–9 (2014).

83. Dahl, E. Some aspects of the ecology and zonation of the fauna on sandy beaches. *Oikos* **4**, 1–27 (1953).

84. Barnes, D. K. A. Ecology of tropical hermit crabs at Quirimba Island, Mozambique: distribution, abundance and activity. *Mar. Ecol. Prog. Ser.* **154**, 133–142 (1997).

85.   Branch, G. M. & Cherry, M. I. Activity rhythms of the pulmonate limpet *Siphonaria capensis* Q. & G. as an adaptation to osmotic stress, predation and wave action. *J. Exp. Mar. Biol. Ecol* **87**, 153–168 (1985).

86.   McLachlan, A., Wooldridge, T. & van der Horst, G. Tidal movements of the macrofauna on an exposed sandy beach in South Africa. *J.Zool.,Lond.* **187**, 433–442 (1979).

87.   Lemaitre, R. & McLaughlin, P. A. World Paguroidea & Lomisoidea database. Coenobita Latreille, 1829. *World Regist. Mar. Species* (2019).

88.   Vannini, M. Researches on the coast of Somalia. The shore and the dune of Sar Uanle 7. Field observations on the periodical transdunal migrations of the hermit crab *Coenobita rugosus* Milne Edwards. *Monit. Zool. Ital.* **7**, 145–185 (1976).

89.   Ince, R., Hyndes, G. A., Lavery, P. S. & Vanderklift, M. A. Marine macrophytes directly enhance abundances of sandy beach fauna through provision of food and habitat. *Estuar. Coast. Shelf Sci.* **74**, 77–86 (2007).

90.   Lemaitre, R. & McLaughlin, P. A. World Paguroidea & Lomisoidea database. *Coenobita rugosus* H. Milne Edwards, 1837. *World Regist. Mar. Species* (2019).

91.   Grubb, P. Ecology of terrestrial decapod crustaceans on Aldabra. *Philos. Trans. R. Soc. London B Biol. Sci.* **260**, 411–416 (1971).

92.   Lemaitre, R. & McLaughlin, P. A. World Paguroidea & Lomisoidea database. Coenobita perlatus H. Milne Edwards, 1837. *World Regist. Mar. Species* (2019).

93.   Mahlia, T. M. I. & Iqbal, A. Cost benefits analysis and emission reductions of optimum thickness and air gaps for selected insulation materials for building walls in Maldives. *Energy* **35**, 2242–2250 (2010).

94.   Vannini, M. Researches on the coast of Somalia. The shore and the dune of Sar Uanle 5. Description and rhythmicity of digging behaviour in *Coenobita rugosus* Milne Edwards. *Monit. Zool. Ital. Suppl.* **13**, 233–242 (1975).

95.   Barnwell, F. H. Daily and tidal patterns of activity in individual fiddler crab (genus *Uca*) from the Woods Hole Region. *Biol. Bull.* **130**, 1–17 (1966).

96.   Barnes, D. K. A. Ecology of tropical hermit crabs at Quirimba Island, Mozambique: a novel and locally important food source. *Mar. Ecol. Prog. Ser.* **161**, 299–302 (1997).

97.   Marsden, I. D. Kelp-sandhopper interactions on a sand beach in New Zealand. I. drift composition and distribution. *J. Exp. Mar. Bio. Ecol.* **152**, 61–74 (1991).

98.   Gibson, R. N. Go with the flow: tidal migration in marine animals. *Hydrobiologia* **503**, 153–161 (2003).

99.   Metcalfe, N. H., Fraser, N. H. C. & Burns, M. D. State-dependent shifts between nocturnal and diurnal activity in salmon. *Proc. R. Soc. B Biol. Sci.* **265**, 1503–1507 (1998).

100.   Metcalfe, N. B. & Steele, G. I. Changing nutritional status causes a shift in the balance of nocturnal to diurnal activity in European Minnows. *Funct. Ecol.* **15**, 304–309 (2001).

101. Orr, M., Zimmer, M., Jelinski, D. E. & Mews, M. Wrack deposition on different beach types: spatial and temporal variation in the pattern of subsidy. *Ecology* **86**, 1496–1507 (2005).

102. Moore, S. L., Gregorio, D., Carreon, M., Weisberg, S. B. & Leecaster, M. K. Composition and distribution of beach debirs in Orange County, California. *Mar. Pollut. Bull.* **42**, 241–245 (2001).

103. Jaramillo, E., de la Huz, R., Duarte, C. & Contreras, H. Algal wrack deposits and macroinfaunal arthropods on sandy beaches of the Chilean coast. *Rev. Chil. Hist. Nat.* **79**, 337–351 (2006).

104. Dugan, J. E., Hubbard, D. M., McCrary, M. D. & Pierson, M. O. The response of macrofauna communities and shorebirds to macrophyte wrack subsidies on exposed sandy beaches of southern California. *Estuar. Coast. Shelf Sci.* **58**, 25–40 (2003).

105. Bell, J. J. Hitching a ride on a hermit crabs home: movement of gastropod shells inhabited by hermit crabs. *Estuar. Coast. Shelf Sci.* **85**, 173–178 (2009).

106. Osenberg, C. W. & Schmitt, R. J. Detecting ecological impacts caused by human activities. in *Detecting ecological impacts: Concepts and applications in coastal habitats* (eds. Osenberg, C. W. & Schmitt, R. J.) 83–108 (Academic Press, 1996). doi:10.1016/B978-012627255-0/50003-3

107. Roberts, C. M. & Hawkins, J. P. Extinction risk in the sea. *Trends Ecol. Evol.* **14**, 241–246 (1999).

108. Brown, A. C. & McLachlan, A. Sandy shore ecosystems and the threats facing them: Some predictions for the year 2025. *Environ. Conserv.* **29**, 62–77 (2002).

109. Houston, J. R. The economic value of beaches — a 2008 update. *Shore & Beach* **81**, 3–11 (2008).

110. Lercari, D. & Defeo, O. Variation of a sandy beach macrobenthic community along a human-induced environmental gradient. *Estuar. Coast. Shelf Sci.* **58**, 17–24 (2003).

111. Benedetti-Cecchi, L. *et al.* Predicting the consequences of anthropogenic disturbance: Large-scale effects of loss of canopy algae on rocky shores. *Mar. Ecol. Prog. Ser.* **214**, 137–150 (2001).

112. Huijbers, C. M., Schlacher, T. A., Schoeman, D. S., Weston, M. A. & Connolly, R. M. Urbanisation alters processing of marine carrion on sandy beaches. *Landsc. Urban Plan.* **119**, 1–8 (2013).

113. Hereward, H. F. R., Gentle, L. K., Ray, N. D. & Sluka, R. D. Ghost crab burrow density at Watamu Marine National Park: An indicator of the impact of urbanisation and associated disturbance? *African J. Mar. Sci.* **39**, 129–133 (2017).

114. Defeo, O. *et al.* Threats to sandy beach ecosystems: a review. *Estuar. Coast. Shelf Sci.* **81**, 1–12 (2009).

115. Brook, S., Grant, A. & Bell, D. Can land crabs be used as a rapid ecosystem evaluation tool? A test using distribution and abundance of several genera from the Seychelles. *Acta Oecologica* **35**, 711–719 (2009).

116. Barboza, F. R. & Defeo, O. Global diversity patterns in sandy beach macrofauna: a biogeographic analysis. *Sci. Rep.* **5**, 1–9 (2015).

117. Belovsky, G. E. & Slade, J. B. Dynamics of two Montana grasshopper populations: relationships among weather, food abundance and intraspecific competition. *Oecologia* **101**, 383–396 (1995).

118. Noriega, R., Schlacher, T. A. & Smeuninx, B. Reductions in ghost crab populations reflect urbanization of beaches and dunes. *J. Coast. Res.* **28**, 123–131 (2012).

119. Lohrer, A. M., Fukui, Y., Wada, K. & Whitlatch, R. B. Structural complexity and vertical zonation of intertidal crabs, with focus on habitat requirements of the invasive asian shore crab, *Hemigrapsus sanguineus* (de Haan). *J. Exp. Mar. Bio. Ecol.* **244**, 203–217 (2000).

120. Schlacher, T. A. *et al.* Human threats to sandy beaches: A meta-analysis of ghost crabs illustrates global anthropogenic impacts. *Estuar. Coast. Shelf Sci.* **169**, 56–73 (2016).

121. Davenport, J. & Davenport, J. L. The impact of tourism and personal leisure transport on coastal environments: a review. *Estuar. Coast. Shelf Sci.* **67**, 280–292 (2006).

122. Gheskiere, T., Vincx, M., Weslawski, J. M., Scapini, F. & Degraer, S. Meiofauna as descriptor of tourism-induced changes at sandy beaches. *Mar. Environ. Res.* **60**, 245–265 (2005).

123. Airoldi, L. *et al.* An ecological perspective on the deployment and design of low-crested and other hard coastal defence structures. *Coast. Eng.* **52**, 1073–1087 (2005).

124. Szabo, K. Terrestrial hermit crabs (Anomura: Coenobitidae) as taphonomic agents in circum-tropical coastal sites. *J. Archaelogical Sci.* **39**, 931–941 (2012).

125. Barnes, D. K. A., Corrie, A., Whittington, M., Carvalho, M. A. & Gell, F. Coastal shellfish resource use in the Quirimba archipelago, Mozambique. *J. Shellfish Res.* **17**, 51–58 (1998).

126. Cabral-Oliveira, J., Mendes, S., Maranhão, P. & Pardal, M. A. Effects of sewage pollution on the structure of rocky shore macroinvertebrate assemblages. *Hydrobiologia* **726**, 271–283 (2014).

127. Saeed, S. & Annandale, D. Tourism and the management of environmental impacts in the Republic of Maldives. *J. Policy Stud.* **7**, 81–88 (1999).

128. Thaman, R. R., Puia, T., Tongabaea, W., Namona, A. & Fong, T. Marine biodiversity and ethnobiodiversity of Bellona (Mungiki) Island, Solomon Islands. *Singap. J. Trop. Geogr.* **31**, 70–84 (2010).

129. Fenberg, P. B. & Roy, K. Ecological and evolutionary consequences of size-selective harvesting: How much do we know? *Mol. Ecol.* **17**, 209–220 (2008).

130. Garcia, S. M. *et al.* Reconsidering the consequences of selective fisheries. *Science (80-.).* **335**, 1045–1048 (2012).

131. Roy, K., Collins, A. G., Becker, B. J., Begovic, E. & Engle, J. M. Anthropogenic impacts and historical decline in body size of rocky intertidal gastropods in southern California. *Ecol. Lett.* **6**, 205–211 (2003).

132. Roome, N. Developing environmental management strategies. *Bus. Strateg. Environ.* **1**, 11–24 (1992).

133. Odum, E. P. The strategy of ecosystem development. *Science (80-.)*. **164**, 262–270 (1969).

134. Turra, A. & Denadai, M. R. Daily activity of four tropical intertidal hermit crabs from southeastern Brazil. *Brazilian J. Biol.* **63**, 537–544 (2003).

135. Kier, G. *et al.* A global assessment of endemism and species richness across island and mainland regions. *PNAS* **106**, 9322–9327 (2009).

136. Laidre, M. E. Ecological relations between hermit crabs and their shell-supplying gastropods: Constrained consumers. *J. Exp. Mar. Bio. Ecol.* **397**, 65–70 (2011).

137. Laidre, M. E. & Vermeij, G. J. A biodiverse housing market in hermit crabs: proposal for a new biodiversity index. *Res. J. Costa Rican Distance Educ. Univ.* **4**, 175–179 (2012).

Printed in the United States
By Bookmasters